Genomics: A Very Short Introduction

VERY SHORT INTRODUCTIONS are for anyone wanting a stimulating and accessible way into a new subject. They are written by experts, and have been translated into more than 45 different languages.

The series began in 1995, and now covers a wide variety of topics in every discipline. The VSI library currently contains over 550 volumes—a Very Short Introduction to everything from Psychology and Philosophy of Science to American History and Relativity—and continues to grow in every subject area.

Very Short Introductions available now:

ACCOUNTING Christopher Nobes
ADOLESCENCE Peter K. Smith
ADVERTISING Winston Fletcher
AFRICAN AMERICAN RELIGION
    Eddie S. Glaude Jr
AFRICAN HISTORY John Parker and
    Richard Rathbone
AFRICAN RELIGIONS Jacob K. Olupona
AGEING Nancy A. Pachana
AGNOSTICISM Robin Le Poidevin
AGRICULTURE Paul Brassley and
    Richard Soffe
ALEXANDER THE GREAT
    Hugh Bowden
ALGEBRA Peter M. Higgins
AMERICAN HISTORY Paul S. Boyer
AMERICAN IMMIGRATION
    David A. Gerber
AMERICAN LEGAL HISTORY
    G. Edward White
AMERICAN POLITICAL HISTORY
    Donald Critchlow
AMERICAN POLITICAL PARTIES
    AND ELECTIONS L. Sandy Maisel
AMERICAN POLITICS
    Richard M. Valelly
THE AMERICAN PRESIDENCY
    Charles O. Jones
THE AMERICAN REVOLUTION
    Robert J. Allison
AMERICAN SLAVERY
    Heather Andrea Williams
THE AMERICAN WEST Stephen Aron
AMERICAN WOMEN'S HISTORY
    Susan Ware

ANAESTHESIA Aidan O'Donnell
ANALYTIC PHILOSOPHY
    Michael Beaney
ANARCHISM Colin Ward
ANCIENT ASSYRIA Karen Radner
ANCIENT EGYPT Ian Shaw
ANCIENT EGYPTIAN ART AND
    ARCHITECTURE Christina Riggs
ANCIENT GREECE Paul Cartledge
THE ANCIENT NEAR EAST
    Amanda H. Podany
ANCIENT PHILOSOPHY Julia Annas
ANCIENT WARFARE
    Harry Sidebottom
ANGELS David Albert Jones
ANGLICANISM Mark Chapman
THE ANGLO-SAXON AGE John Blair
ANIMAL BEHAVIOUR
    Tristram D. Wyatt
THE ANIMAL KINGDOM
    Peter Holland
ANIMAL RIGHTS David DeGrazia
THE ANTARCTIC Klaus Dodds
ANTHROPOCENE Erle C. Ellis
ANTISEMITISM Steven Beller
ANXIETY Daniel Freeman and
    Jason Freeman
APPLIED MATHEMATICS
    Alain Goriely
THE APOCRYPHAL GOSPELS
    Paul Foster
ARCHAEOLOGY Paul Bahn
ARCHITECTURE Andrew Ballantyne
ARISTOCRACY William Doyle
ARISTOTLE Jonathan Barnes

STEM CELLS Jonathan Slack
STRUCTURAL ENGINEERING
    David Blockley
STUART BRITAIN John Morrill
SUPERCONDUCTIVITY
    Stephen Blundell
SYMMETRY Ian Stewart
TAXATION Stephen Smith
TEETH Peter S. Ungar
TELESCOPES Geoff Cottrell
TERRORISM Charles Townshend
THEATRE Marvin Carlson
THEOLOGY David F. Ford
THINKING AND REASONING
    Jonathan St B. T. Evans
THOMAS AQUINAS Fergus Kerr
THOUGHT Tim Bayne
TIBETAN BUDDHISM
    Matthew T. Kapstein
TOCQUEVILLE Harvey C. Mansfield
TRAGEDY Adrian Poole
TRANSLATION Matthew Reynolds
THE TROJAN WAR Eric H. Cline
TRUST Katherine Hawley
THE TUDORS John Guy
TWENTIETH-CENTURY
    BRITAIN Kenneth O. Morgan
THE UNITED NATIONS
    Jussi M. Hanhimäki

THE U.S. CONGRESS Donald A. Ritchie
THE U.S. SUPREME COURT
    Linda Greenhouse
UTILITARIANISM
    Katarzyna de Lazari-Radek and
    Peter Singer
UNIVERSITIES AND COLLEGES
    David Palfreyman and Paul Temple
UTOPIANISM Lyman Tower Sargent
VETERINARY SCIENCE James Yeates
THE VIKINGS Julian Richards
VIRUSES Dorothy H. Crawford
VOLTAIRE Nicholas Cronk
WAR AND TECHNOLOGY Alex Roland
WATER John Finney
WEATHER Storm Dunlop
THE WELFARE STATE David Garland
WILLIAM SHAKESPEARE
    Stanley Wells
WITCHCRAFT Malcolm Gaskill
WITTGENSTEIN A. C. Grayling
WORK Stephen Fineman
WORLD MUSIC Philip Bohlman
THE WORLD TRADE
    ORGANIZATION Amrita Narlikar
WORLD WAR II Gerhard L. Weinberg
WRITING AND SCRIPT
    Andrew Robinson
ZIONISM Michael Stanislawski

Available soon:

GEOPHYSICS William Lowrie
CRIMINOLOGY Tim Newburn
SEXUAL SELECTION Marlene Zuk and
    Leigh W. Simmons

DEMOGRAPHY Sarah Harper
GEOLOGY Jan Zalasiewicz

For more information visit our website

www.oup.com/vsi/

John Archibald

GENOMICS

A Very Short Introduction

OXFORD
UNIVERSITY PRESS

OXFORD

UNIVERSITY PRESS

Great Clarendon Street, Oxford, OX2 6DP,
United Kingdom

Oxford University Press is a department of the University of Oxford.
It furthers the University's objective of excellence in research, scholarship,
and education by publishing worldwide. Oxford is a registered trade mark of
Oxford University Press in the UK and in certain other countries

Published in the United States of America by Oxford University Press
198 Madison Avenue, New York, NY 10016, United States of America

British Library Cataloguing in Publication Data
Data available

Library of Congress Control Number: 2017959740

ISBN 978–0–19–878620–7

Printed and bound by
CPI Group (UK) Ltd, Croydon, CR0 4YY

Contents

Preface

Genomics has transformed the biological sciences. From epidemiology and medicine to evolution and forensics, the ability to determine an organism's complete genetic makeup has changed the way science is done and the questions that can be asked of it. Far and away the most celebrated achievement of genomics is the Human Genome Project, a technologically challenging endeavour that took thousands of scientists around the world thirteen years and ~US$3 billion to complete. In 2000, American President William Clinton referred to the resulting genome sequence as 'the most important, most wondrous map ever produced by humankind.' Important though it was, this 'map' was a low-resolution first pass—a beginning not an endpoint. As of this writing, thousands of human genomes have been sequenced, the primary goals being to better understand our biology in health and disease, and to 'personalize' medicine. Sequencing a human genome now takes only a few days and costs as little as US$1,000. The genomes of simple bacteria and viruses can be sequenced in a matter of hours on a device that fits in the palm of your hand. The information is being used in ways unimaginable only a few years ago.

The term *genomics* can mean different things to different people. It is in one sense a collection of experimental methods used to analyse the sequence and structure of an organism's genome.

Genomics is also a mature, far-reaching branch of science whose subject is 'the genome' and the genes contained therein. This book explores the science of genomics and its rapidly expanding toolbox. The goal is to provide the reader with a broad overview of the topic, including the molecular biology upon which genomics is based; examples of the scientific questions it is used to address; how it impacts our daily lives; and how it is likely to do so in the future. As we shall see, genomics is a fast-paced field, one that continues to push not only technological boundaries but social, legal, and ethical ones as well. Indeed, DNA sequencing techniques are evolving so rapidly that methods developed within the last decade have already become obsolete. Exploration of the topic requires a certain amount of technical detail, but I have endeavoured to keep such information to a minimum. Readers should also feel free to jump to wherever their interests take them; lists of further reading and references for each chapter appear at the end of the book for those wishing to dig deeper into specific topics.

Acknowledgements

Modern science is so fast-paced and has become so specialized that it is hard to keep on top of one's own discipline, let alone a field as broad as genomics. As a technology-driven area of research, advances in genomics happen particularly rapidly. In writing this book, I have benefited from the work of a great many scientists. I am particularly indebted to the following individuals, listed alphabetically, for their time and effort in providing comments on some or all of the chapters herein: Karen Bedard, Zhenyu Cheng, André Comeau, Jennifer Corcoran, Bruce Curtis, Jan de Vries, Matt Field, Peter Holland, Philip Hugenholtz, Jon Jerlström-Hultqvist, Nicole King, Rasmus Nielsen, Iñaki Ruiz Trillo, Beth Shapiro, and David Walsh. I would also like to thank Latha Menon and Jenny Nugée of Oxford University Press for providing expert editorial assistance along the way.

List of illustrations

Genomics

Common abbreviations

BAC	bacterial artificial chromosome
bp	base pairs
cDNA	complementary DNA
DNA	deoxyribonucleic acid
EST	expressed sequence tag
HGT	horizontal gene transfer
Kbp	kilobase pairs
Mbp	megabase pairs
nm	nanometres
PCR	polymerase chain reaction
RNA	ribonucleic acid
mRNA	messenger RNA
RNA-seq	RNA sequencing (whole transcriptome shotgun sequencing)
rRNA	ribosomal RNA
SNP	single-nucleotide polymorphism

Chapter 1
What is genomics?

Elucidating the nature of heredity was one of the greatest
triumphs of 20th-century science. First discovered in 1869 by a
Swiss doctor named Friedrich Miescher, DNA—deoxyribonucleic
acid—is often described as the 'master molecule' of life: it is
what genes are made of, the genetic material through which
organisms pass on physical traits and innate behaviours to their
offspring. Structurally and chemically speaking, DNA is a
relatively simple molecule, but it is capable of storing vast
quantities of information. Biologists use the term *genome* to refer
to the sum total of DNA inside the single or multiple cells that
make up an organism. The genome includes all of the organism's
genes as well as the so-called 'non-coding' DNA. The amount of
non-coding DNA (sometimes called 'junk' DNA) in a genome can
differ from species to species. It is also important to note that
complex organisms often have more than one genome: DNA can
be contained in two or more distinct compartments within the
cell. Human beings, for example, have two genomes, one residing
in the cell nucleus and another in our cellular 'powerhouse',
the mitochondrion. The size and structure of genomes varies
significantly depending on the type of organism in which
they reside.

The biology of cells

The fundamental unit of life on Earth is the cell and there are two main types: prokaryotes (pronounced: pro-care-ee-oats) and eukaryotes (you-care-ee-oats). Prokaryotes are microscopic, single-celled organisms that include the myriad of bacteria known to science: the *Escherichia coli* in our gut and the plague-causing *Yersinia pestis*; the *Streptococcus* that contribute to tooth decay and the *Staphylococcus* that cause skin infections; the *Lactobacillus* we use to make yoghurt; the cyanobacteria that photosynthesize in the ocean; and the nitrogen-fixing bacteria that live symbiotically with the roots of plants. All of these bacteria and many more are prokaryotic organisms.

Prokaryotes also include the *Archaea*, or Archaebacteria, a lesser-known group of single-celled organisms that often inhabit extreme environments. Archaea can live at temperatures of more than 100°C in hot springs and hydrothermal vents at the bottom of the ocean, while others thrive in high-salt environments such as solar salterns, which can be up to ten times more salty than seawater. Still other archaea are 'acidophiles': they have been found inhabiting sulfuric pools and acid mine drainages at pH levels approaching 0 (on a scale of 0 to 14).

Unlike prokaryotes, which have little in the way of internal structure, eukaryotic cells are characterized by the presence of numerous membrane-bound internal compartments (see Figure 1). These include the nucleus, in which the main genome of the cell resides, as well as the mitochondrion, an important organelle that is the site of adenosine triphosphate (ATP) production. In the case of photosynthetic eukaryotes, a light-harvesting chloroplast is also present. Mitochondria and chloroplasts both contain genomes, a legacy of the fact that they evolved from once free-living bacteria that came to reside inside a eukaryotic host cell. Eukaryotes are further characterized by the

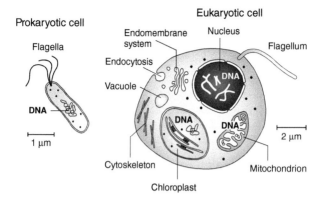

Prokaryotic cell

Eukaryotic cell

Flagella

Endomembrane system

Nucleus

Flagellum

Endocytosis

Vacuole

DNA

DNA

DNA

1 μm

2 μm

Cytoskeleton

Mitochondrion

Chloroplast

1. **Prokaryotic and eukaryotic cells.**

presence of an elaborate 'cytoskeleton', a protein-based internal scaffolding system that allows them to change their shape and grow to a much larger size than prokaryotic cells.

Essentially all organisms visible to the naked eye are eukaryotes—animals, plants, fungi, kelp, and so on—as are a wide array of microscopic, single-celled creatures such as pond algae, ocean phytoplankton, soil amoebae, and pathogens such as the causative agent of malaria, *Plasmodium*. The nuclear genomes of eukaryotes consist of multiple linear DNA molecules, whereas prokaryotic genomes, as well those of organelles, are typically circular in nature. Various other differences between eukaryotic and prokaryotic cells will be considered as we move through the chapters that follow.

Note that I have not included viruses in this overview of cell biology. This is because they are neither prokaryotes nor eukaryotes. In fact, many biologists do not consider viruses to be alive at all because they are dependent on the machinery of a cell in order to replicate (viruses of both prokaryotes or eukaryotes are known). This is not to belittle their importance: there are many more viruses on Earth than there are bacteria (and there are a *lot* of bacteria), and viruses are astonishingly diverse in form and function. Viruses are also unique in that some of them contain a ribonucleic acid (RNA)-based genome, unlike cellular organisms, which exclusively use DNA as their genetic material. We will explore the diversity of viruses and their genomes in Chapter 6—it is an exciting area of research, one that is enjoying a resurgence of interest thanks to modern DNA sequencing techniques.

Molecular biology

The iconic double-helical structure of DNA holds the secret to understanding not only how genes are passed on to offspring but also how they 'direct' the synthesis of proteins, the main building blocks from which all cells are constructed. It is useful to think of DNA as a spiralling ladder. The steps of the ladder are built from nucleotides, which contain a negatively charged phosphate on the outside, a so-called base on the inside, and a sugar linking the two. There are four bases in DNA—adenine (A), cytosine (C), guanine (G), and thymine (T)—and their specific chemical characteristics govern how the two strands of the double helix interact: A pairs with T, C pairs with G. These A-T and C-G 'base pairs' (bp) make up the internal rungs of the DNA ladder (see Figure 2).

DNA replication involves pulling the two strands of the double helix apart. Once separated, the cell's protein machinery then uses each of the single strands as a template for the sequential addition of complementary nucleotides: A residues are added across from Ts, and Cs are inserted across from Gs. The end result is two DNA double helices, both identical to the original. DNA replication is extremely accurate but it is not perfect; it is during this process that most mutations in DNA arise. The concept of nucleotide incorporation is one that we will revisit in more detail in the next chapter, as it forms the foundation of most of the techniques used to sequence DNA in the laboratory. The complementary nature of the two strands of DNA means that only one strand needs to be

4

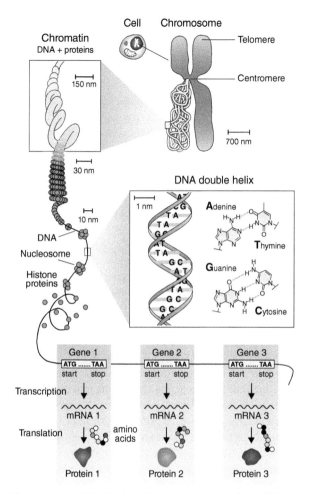

2. **Chromosomes, DNA, RNA, and protein.** Diagram shows (i) the packaging of DNA and protein into chromatin and chromosomes, which are located in the nucleus of a eukaryotic cell; (ii) the double-helical structure of DNA, including the four chemical 'building blocks' from which it is made; and (iii) how DNA gives rise to a chemical intermediate called RNA, and how the information in the RNA is used as a template for the synthesis of proteins.

sequenced—knowing the order of bases on one strand means that the sequence of the other strand can be inferred.

The mechanism by which genes mediate the synthesis of proteins also involves separation of the double helix. In this case, one of the two DNA strands serves as a template for the production of RNA, a nucleic acid whose chemical composition is similar to, but distinct from, DNA. The sugar in RNA is ribose, not deoxyribose as in DNA, and the ribonucleotide bases are A, C, G, and U (uracil). U is chemically similar to T and thus pairs with A. Depending on the RNA produced, it may function exclusively as a single-stranded molecule or it may fold up into a complex three-dimensional structure comprised of alternating single- and double-stranded regions.

RNA molecules serve as a critical link between the genes that reside in the DNA and the proteins they specify. While DNA and RNA are linear chains comprised of nucleotides, proteins are polymers of amino acids. There are twenty different types of amino acids in naturally occurring proteins, each 'hard-wired' into the so-called universal genetic code. The genetic code used by all cells (prokaryotic and eukaryotic) is a triplet code, which is to say that each particular amino acid is 'encoded' by a three-base 'codon'. These codons appear one after the other in the gene sequence of the DNA and are converted into RNA by a process called *transcription*. This so-called 'messenger RNA' (mRNA) is then used as a template for the synthesis of proteins: during *translation*, amino acids are linked together in a linear fashion in the precise order specified by the codons in the gene (see Figure 2).

Protein synthesis is a remarkably complex process, one that involves dozens of protein and RNA molecules. For our purposes, it is sufficient to appreciate that because the same genetic code is used for protein synthesis in the vast majority of organisms that have been examined, we can predict with high confidence the amino acid sequence of any given protein from the DNA sequence of its gene. Simply put, protein-coding genes are long strings of three-nucleotide codons, and with the aid of a computer we can search all of the chemical letters that make up a given genome and identify the genes within it. (It should be noted that a handful of instances of 'codon reassignment' have been documented—i.e. cases in which certain triplet codons within an organism's genome are used to specify a different amino acid than in the canonical code. These examples are nevertheless extremely rare and the genetic code is thus generally said to be 'universal'; its complex and somewhat arbitrary nature speaks strongly to the common ancestry of all life on Earth.)

Genomes by numbers

To completely sequence an organism's genome is to determine the precise order of all of the As, Cs, Gs, and Ts in its DNA. Genome size varies tremendously from species to species, and the larger the genome of interest the greater the challenges associated with sequencing it. Not surprisingly, genomics as a discipline began by studying the small genomes of prokaryotes and viruses. The first genome was sequenced in 1976—it was that of an RNA virus named MS2, a 'bacteriophage' that infects *E. coli* and related bacteria. The MS2 genome is about as simple as a genome can get: it is ~3,600 bp long and contains only four protein-coding genes. The first DNA-based genome was sequenced shortly thereafter, that of bacteriophage phiX174. This virus has a long history in molecular biology research and is still used today. For example, due to its small size (~5,400 bp and eleven protein-coding genes), the phiX174 genome is often used as a positive control in genomics labs to ensure that the sequencing instruments and bioinformatic tools are functioning properly.

The bacterium *E. coli*, a lab workhorse in both basic and applied research settings, has a genome of ~4.6 million bp and has ~4,300 protein-coding genes (the genomes of hundreds of different *E. coli* strains have now been sequenced; the substantial variation seen between them gives rise to important differences in their pathogenicity). Stretched out in a line, the *E. coli* genome is about 1.5 millimetres long, which is remarkable given that the DNA double helix is a mere 2 nanometres in diameter and an individual *E. coli* cell is only ~2 micrometres in length. *E. coli* and other bacteria can also harbour 'plasmids', tiny circular DNA molecules that are physically distinct from the cell's main genome. Plasmids are typically <10,000 bp in size and contain only a few to a few dozen genes. In nature, plasmids are frequently transferred from cell to cell; in the lab they are easily manipulated and thus used extensively in genetic engineering.

The human genome is made up of ~3.2 *billion* bp of DNA. The average human chromosome contains ~5 centimetres of DNA and considering all twenty-three pairs of chromosomes, there are ~2 *metres* of DNA packed into the nucleus of each of our cells. These numbers highlight the extensive compaction that must take place in order to cram the DNA inside the cell (see Figure 2). They also speak to the difficulties encountered when trying to sequence a genome and identify the particular stretches of DNA that correspond to genes.

We still do not know how many genes are in the human genome. Current estimates are between 19,000 and 25,000 but no one knows for sure. There are various reasons for this. Much of the uncertainty revolves around the fact that only ~1 per cent of the human genome corresponds to protein-coding genes, and the genes themselves are often broken up into pieces. In prokaryotic genomes, genes are contiguous, uninterrupted strings of amino acid-specifying codons, and the process of gene finding is usually straightforward. In eukaryotes, however, the codon-containing regions of genes, the 'exons', are interrupted by stretches of non-coding DNA—'introns'. The introns are transcribed into RNA but are removed ('spliced') before the molecule is used as a template for protein synthesis. Human genes often have dozens of introns, each of which can be tens of thousands of nucleotides long. Distinguishing exons from introns and other forms of non-coding DNA is thus challenging. Molecular biologists have developed various techniques for sifting through vast quantities of genome sequence data and honing in on just the protein-specifying codons, but the tools are still far from perfect and what works for one organism's genome may not work for another's.

A related issue is the fact that large regions of the human genome have yet to be sequenced due to technical challenges associated with sequencing certain types of DNA, such as repetitive strings of nucleotides. Given the nature of these sequences, it is unlikely that these missing regions are gene-rich, but it is formally possible that there are still genes lurking in such portions of the genome. Efforts to improve the quality of the human genome sequence are ongoing, taking advantage of the latest tools in DNA sequencing and analysis.

It is also important to recognize that the precise definition of a gene can vary. Although genome-sequencing efforts tend to focus on the protein-coding genes, all cells possess a variety of non-coding RNA genes whose functional end product is not protein but the RNA molecule itself. Some of these non-coding RNAs have been known for a long time. These include ribosomal RNAs, which are RNA components of the protein-synthesizing *ribosome*, and transfer RNAs whose job is to bring individual amino acids to the site of protein chain elongation. One of the revelations of the past decade of genomics research has been the realization that hidden among the protein-coding genes are many hundreds of small RNA-encoding genes whose products exhibit a level of functional diversity greatly exceeding what was originally thought possible. These include microRNAs, a class of small regulatory RNAs that bind to mRNAs and in so doing modulate the expression of protein-coding genes. By influencing when such genes are turned on and off, microRNAs have the potential to influence core cellular processes such as DNA replication and cell division.

There is still much to learn about the size and complexity of the 'black box' of small RNA biology, and again technological advances have allowed researchers to head in exciting new directions. The concept of 'the gene' continues to evolve.

Why sequence a genome?

Before turning our attention to the nuts and bolts of how a genome is sequenced, let's consider why there is so much interest and excitement in doing so. Of what use is an organism's genome sequence? Why is it so important to have a list of all of its genes? DNA is sometimes referred to as a 'recipe' for life, the genome as the 'parts list' of the cell. These analogies are far from perfect, but it is certainly the case that without a complete gene inventory, an organism's biology can never be fully understood. In isolation, a genome sequence does not tell us everything we need to know—far from it. There are still many genes in humans and other organisms whose functions have yet to be described. Knowing the sequence of a gene (and the protein it encodes) does not necessarily tell us anything about its job(s) in the cell, and genes do not work in isolation. Genes and the environment interact in complex ways; both play important roles in the life of an organism. But a complete genome sequence is an invaluable—and ever more attainable—starting point from which researchers can obtain information about the functions of genes and proteins through laboratory experimentation.

As important as having a list of an organism's genes is knowledge of when the genes are active or inactive during its life history. In complex multicellular species, only a fraction of the genes in the genome are switched on (i.e. transcribed into mRNA and used to make proteins) at any given time. It is the combination of particular subsets of proteins in different cells at the right time that lead to cellular and tissue differentiation. Even in single-celled microbes, gene expression patterns change in response to their surrounding environment: changes in light intensity, nutrient concentrations, abundance of viruses, etc. If examined systematically, patterns of gene expression can provide useful clues as to what the functions of particular genes and proteins might be. An important example is the study of human diseases such as cancer, which often develops when certain genes are mutated and/or inappropriately turned on or off.

As we shall see in Chapter 5, complete genome sequences are being used to address fundamental questions in evolutionary biology across a wide range of timescales. These include the study of early human migration patterns and how we humans differ from our closest non-human relatives, the now-extinct Neanderthals. Genomic data are also being used to investigate how multicellular animals and plants evolved, and further still back in time, how the first eukaryotic cells evolved from prokaryotic antecedents. And as will be discussed in several chapters, a powerful approach called 'metagenomics' (or environmental genomics) allows researchers to move beyond the analysis of one genome at a time to study complex mixtures of DNA extracted directly from the environment, including soil, water, and even air samples. The human gut is an area of particularly intense metagenomic investigation. DNA sequence data taken from such samples can tell us a lot about the microbial communities within them—we can now ask 'who's there?' as well as 'what are they doing?' without ever having used a microscope. Underlying each of these diverse research foci is a common set of DNA sequencing and genome assembly tools. And because DNA is DNA, no matter what the source, it doesn't take long for technological developments in one area of genomics to spread to others.

Chapter 2
How to read the book of life

For all its biological importance, DNA is a fragile molecule. Its long, threadlike structure makes it difficult to study in its natural state. In living cells, DNA rarely exists in isolation; it is typically bound tightly to proteins (see Figure 2), proteins that serve to compact it and to regulate the process of gene expression. Consequently, the act of purifying (or 'extracting') DNA from the multitude of proteins, lipids, and other macromolecules inside the cell requires the use of harsh chemicals and subjects it to physical damage. While it is possible to estimate the size of intact chromosomes in the lab, modern DNA sequencing instruments typically collect at most 1,000 base pairs of nucleotide sequence per DNA fragment. To sequence a genome, therefore, scientists must first generate thousands to millions of individual short sequence 'reads' (the bigger the genome the more reads are required), which are then compared to one another *in silico* and used to reconstruct (or 'assemble') a picture of an organism's DNA as it exists in the cell. In this chapter we will cover the sequencing techniques themselves, and in Chapter 3 focus on the tools used for genome assembly and gene finding. Technological advances in both of these areas have brought the field of genomics to where it is today.

Protein and RNA first

Despite its current ubiquity, DNA sequencing was developed well after RNA and protein sequencing. There are several reasons for this, the most important being that initially protein and RNA molecules were more easily purified than DNA, and thus more amenable to experimentation. The first protein to have its amino acids 'read' to completion was the small hormone insulin, which was sequenced in the mid-1950s by the British biochemist Frederick Sanger. Sanger's protein-sequencing methods will not be discussed here, but his seminal contributions to biochemistry and molecular biology run throughout this book and are worthy of repeated mention (Sanger received the Nobel Prize in Chemistry twice, in 1958 and 1980, for the development of protein and DNA sequencing, respectively).

Methods for RNA sequencing came roughly a decade later. By necessity, the initial focus was on small, easily purified molecules such as transfer RNAs. In 1965, a seventy-seven-nucleotide transfer RNA was purified from yeast cells and completely sequenced by the American biochemist and Nobel laureate Robert Holley. Throughout the 1960s, Sanger had experimented with new methods for the study of small RNAs, and it was the American microbiologist Carl Woese who built upon them with the goal of tackling much larger RNA molecules. Woese's scientific interests revolved around the ribosome, the large, super-abundant RNA-protein complex responsible for protein synthesis; he was the first to collect nucleic acid sequences for the purposes of using them as evolutionary chronometers.

The 'RNA cataloguing' technique of Woese and colleagues involved growing cell cultures in the presence of radioactive phosphate. As cells synthesize new ribosomal components (which they must do constantly), the phosphate is incorporated into ribonucleotides and, eventually, ribosomal RNAs (rRNAs). Radioactively labelled rRNAs could then be purified by gel electrophoresis, a technique that involves separating molecules by size by passing them through a porous, gelatinous substance in the presence of an electric field. (The use of 'labels' is a common feature in molecular research; as we shall see, a variety of different labels are used in DNA sequencing but the purpose is always the same: to be able to tag newly synthesized molecules so that they can be traced during the course of an experiment.)

To obtain actual sequence information, the purified and radioactive rRNAs were chopped into smaller pieces with specific RNA-degrading enzymes (e.g. RNase T_1, which cuts after G ribonucleotides). The samples were again separated according to size, and after multiple rounds of digestion and separation, what remained were diagnostic patterns of black spots on an X-ray film. Each spot corresponds to an RNA fragment only five or six nucleotides long. With effort, the exact nucleotide sequence of each fragment could be 'puzzled out' by considering its migration during electrophoresis and the enzymes that gave rise to it. In Woese's day, each RNA sequence became an entry in the lab RNA catalogue, analyses of which eventually led to the discovery of the third domain of life, the Archaea.

DNA sequencing

RNA-based techniques for molecular sequencing are laborious in the extreme, and in the late 1970s they were quickly supplanted by two DNA sequencing methods developed independently by Sanger and the American molecular biologist Walter Gilbert. Gilbert's 'chemical sequencing' approach is still useful in certain research applications, but here we will focus on what became known as Sanger sequencing, as this is the one that, for various reasons, took hold within the research community. It is still widely used today, albeit with modern instruments and reagents.

Sanger's method typically starts with the DNA fragment to be sequenced inserted into a circular plasmid molecule to which a short DNA sequencing 'primer' has been bound. These DNAs are then mixed with a DNA polymerase enzyme, the same enzyme that in nature replicates DNA in the cell. To this mixture, each of the four building blocks of DNA is added: these are the deoxynucleotides dATP, dCTP, dGTP, and dTTP, one of which is tagged with radioactivity. Critical to the success of the experiment is the addition of dideoxynucleotides (ddNTPs) as well. These ddNTPs act as chain terminators: they are chemically modified in such a way that they can be incorporated into DNA by the polymerase but cannot be used for further chain extension. Classical Sanger sequencing uses four separate reaction vessels for each DNA sample to be sequenced, which differ only in the identity of the ddNTP added (each tube contains a different DNA chain terminator).

In developing his procedure, Sanger's brilliance was to tweak the relative amounts of dNTPs and ddNTPs present in the reaction tubes such that ddNTP incorporation, and thus chain termination, happens only rarely. In the A tube, each time a T residue is encountered on the DNA template the polymerase enzyme will add a normal dATP most of the time. This is because A pairs with T and the tube contains more dATP molecules than ddATPs. Chain elongation can continue. Upon reaching the next T in the template, chain termination is again unlikely, but possible. The end result is an A tube full of radioactive DNA fragments of multiple sizes, each indicating the position of a T residue in the DNA template. The reactions taking place in the three additional tubes are identical to the first, except that the newly synthesized DNA fragments are terminated by ddCTP, ddGTP, and ddTTP.

The final step is to use gel electrophoresis to separate the DNA fragments in each of the four tubes side by side. This results in a DNA sequencing 'ladder', which (because of the use of radioactive dNTPs) can be visualized by exposing a piece of X-ray film to the gel. By analysing the migration patterns of the rungs on the ladder, the sequence of the original DNA molecule can be inferred.

Sanger's original version of dideoxy chain-termination sequencing was a quantum leap beyond RNA cataloguing. Because it involved DNA synthesis in a test tube, it sidestepped the need to obtain large quantities of starting material (limiting which nucleic acids could be sequenced). It was nevertheless still time-consuming and subject to human error (I speak from experience). The gels needed for DNA separation are messy to prepare, and the nucleotide sequences are read off the X-ray films by eye. Typically only 150–200 bp of sequence can be read per reaction, meaning that for longer DNA fragments, multiple-sequencing reactions using different DNA primers need to be performed in an iterative fashion.

In response to ever-increasing demand, Sanger sequencing has been refined considerably over the years. In the mid-1980s, experimentation with the use of fluorescent (as opposed to radioactive) labels led to the development of 'automated' DNA sequencers. With fluorescence, chemically distinct tags can be used for each of the four ddNTPs, meaning that all possible chain termination products can be synthesized in the same tube and run in the same lane of a sequencing gel (see Figure 3). The use of heat-stable DNA polymerases, like those used in the 'polymerase chain reaction' (a technique for amplifying defined stretches of DNA from minute quantities of starting material; see Chapter 5), helped to further automate and speed up the process. Modern Sanger sequencing instruments separate the labelled DNA fragments by passing them through fine capillary tubes; the machine (not a human being) reads and records the DNA sequence by shining a laser on the sample as it runs through the bottom of the tube and detecting which of the four fluorescent tags are present.

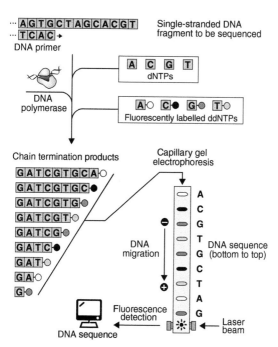

Single-stranded DNA fragment to be sequenced

DNA primer

DNA polymerase

dNTPs

Fluorescently labelled ddNTPs

Chain termination products

Capillary gel electrophoresis

DNA migration

DNA sequence (bottom to top)

Fluorescence detection

DNA sequence

Laser beam

3. **Di-deoxy chain terminator (Sanger) sequencing (automated procedure).**

DNA sequencing technologies have taken several more quantum leaps over the past decade. Before moving on, however, it is worth noting that there is still a niche for Sanger's method. Using current instrumentation, it is still the gold standard in terms of data quality, and it produces very respectable read lengths (800–1,000 bp on average). Automated Sanger sequencing is also the most cost-effective method if one's needs are modest (e.g. if fewer than several dozen short DNA molecules need to be sequenced). Note too that we will forgo discussion of genome assembly until Chapter 3, after exploring the various 'next-generation' sequencing technologies. By and large, the challenges associated with stitching a genome together from individual sequencing reads are the same, regardless of how those reads are generated.

Next-generation DNA sequencing

At its peak, the government-funded Human Genome Project relied on the sequencing capacities of more than a dozen universities and research centres in the US, the UK, France, Germany, Japan, and China. Each was filled with dozens of state-of-the-art Sanger sequencing instruments, each sequencing 96 or 384 DNA samples in parallel, twenty-four hours a day, seven days a week. Data quality was high and the task of sequencing the human genome was completed ahead of schedule. But it was expensive—US\$3 billion expensive. The privately funded project, carried out by American Craig Venter's Celera Corporation, began much later and produced a genome sequence of roughly the same quality with the same instrumentation for ~US\$300 million. (In doing so, Celera took advantage of years of preliminary data generated by the public project; see Chapter 3 for discussion of the different strategies employed.) Having been continuously tweaked and streamlined, Sanger sequencing had by the late 1990s approached the limit of data generation per instrument per unit time.

What emerged in the mid-2000s, shortly after the public and private human genome projects were declared finished, was a sequencing technique capable of generating orders of magnitude more data at a fraction of the cost. The approach was called 'pyrosequencing', and what it shares with all of the so-called 'next-generation' technologies that have come after it is massive parallelization—the ability to carry out many thousands to millions of individual sequencing reactions all at the same time on a single machine. Remarkably, while most next-generation sequencing instruments are still quite large (roughly the same size as modern Sanger sequencers), the reactions themselves take

18

place on the surface of a rectangular glass slide only a few centimetres across.

In contrast to Sanger sequencing, which uses gel electrophoresis to 'read' the products of chain-termination reactions performed in advance, pyrosequencing records DNA synthesis as it happens in real time. The chemistry, developed by Swedish biochemist Pål Nyrén, is built around detection of the pyrophosphate (PP_i) released each time a polymerase adds a dNTP to a DNA template. This is achieved through the use of a multi-enzyme chemical cascade that uses the PP_i to generate a burst of bioluminescence.

The first commercially available pyrosequencing instrument—and indeed the very first next-generation sequencer—came on line in 2005. It was produced by 454 Life Sciences, a company founded by the American scientist-entrepreneur Jonathan Rothberg (Nyrén and colleagues licensed their technology to 454 Life Sciences). The key to its ability to massively parallelize DNA sequencing is use of a 'picotiter' plate containing many thousands of individual reaction wells, each ~25 micrometres in diameter. Each well contains a single bead, bound to which are thousands of identical copies of a single-stranded DNA template, amplified from a single 'seed' molecule (this amplification step is necessary in order to bring the bioluminescence signal generated in each well up to detectable levels). The DNA contained in each well is primed and ready to be extended when dNTPs and the appropriate enzymes are added.

Because pyrosequencing uses a single 'signal' for detection of nucleotide incorporation (i.e. light), dNTPs must be presented to the DNA template in a systematic manner. This is achieved using a microfluidics system, which flows the four dNTPs across the picotiter plate one at a time and in a defined order. A powerful camera is focused on the plate and records whether or not light emission (and thus dNTP incorporation) has taken place in each well.

An important limitation of pyrosequencing is the so-called 'homopolymer' problem: when more than a few identical nucleotides are encountered on a DNA template, the instrument has difficulty inferring the precise number of dNTP incorporations that have occurred. For example, a string of eight consecutive As in the template will result in the incorporation of eight dTTPs, but reliably distinguishing between the bioluminescence signals generated by seven, eight, or nine incorporations is difficult. The longer the homopolymer tract the bigger the problem becomes.

Homopolymers aside, technological refinements have increased the density of wells per picotiter plate up to ~1 million and average read lengths of >500 bp can be generated—a significant improvement over the ~100 bp reads produced by the first-generation machines. The most advanced pyrosequencing instruments are capable of generating ~500 million bp of sequence data during the course of a ten-hour run. The tremendous potential of pyrosequencing was demonstrated in 2006 with the sequencing of one million bp of Neanderthal DNA (see Chapter 5), and by completely sequencing the genome of the Nobel laureate James Watson in 2007 (see Chapter 4). Watson's genome was sequenced in two months at a cost of ~US$2 million, a small fraction of the cost if using Sanger sequencing.

Semiconductor sequencing

A testament to the rapid evolution of DNA sequencing methods is that for all its potential, 454 pyrosequencing has come and gone in the span of a decade, no longer deemed competitive with various other emerging platforms. (Rothberg's 454 Life Sciences was acquired by Roche Diagnostics in 2007, which stopped supporting its pyrosequencing instruments in 2016.) Before moving further afield, let's briefly explore a new technology from a company called Ion Torrent (also the brainchild of Rothberg), whose semiconductor sequencing is similar to pyrosequencing in many respects but nevertheless highly distinct in terms of sequence detection.

With Ion Torrent's semiconductor sequencing, there is no enzyme-driven chemical cascade leading to a burst of light. Rather, dNTP incorporations are detected by 'sensing' the hydrogen ions (H^+) generated during DNA synthesis. Nucleotides are added to the flow cell sequentially, as in pyrosequencing. Each dNTP incorporation results in the release of a single H^+, and, remarkably, millions of sensors contained in an ion-sensitive chip beneath the wells of the picotiter plate detect the resulting changes in acidity (pH). The approach is described as 'non-optical' semiconductor DNA sequencing—in essence, tiny pH meters convert chemical information (i.e. H^+) into digital sequence data, in the same way that the semiconductor chip in a digital camera converts light into digital information.

Sequence read lengths with Ion Torrent are impressive (currently up to ~400 bp) and the device is much faster than other next-generation platforms (only a few hours run time). As a proof of principle, Ion Torrent was used to sequence a human genome in 2011 (fittingly, that of Gordon Moore, co-founder of the electronics giant Intel Corporation and coiner of 'Moore's law', which describes the doubling of computer chip performance every eighteen months). Overall, however, sequence output per instrument run is currently modest and the technology is sensitive to the homopolymer problem described earlier for pyrosequencing. At present, the Ion Torrent platform is marketed towards rapid clinical sequencing and related applications (e.g. detecting cancer mutations) rather than sequencing genomes from scratch.

Reversible chain-termination sequencing

One of the most widely utilized next-generation DNA sequencing platforms is Illumina. Like Sanger and pyrosequencing, Illumina employs a DNA polymerase-based, 'sequencing-by-synthesis' approach. Where it differs is in its use of so-called 'reversible terminator' technology. As the name suggests, it utilizes modified dNTPs that act as chain terminators: they can be incorporated into DNA but cannot be extended until a specific chemical 'block' has been removed.

Illumina determines the sequence of clusters of DNA randomly attached to the surface of a solid glass slide (each cluster contains ~1,000 identical DNA molecules) (see Figure 4). As the sites of

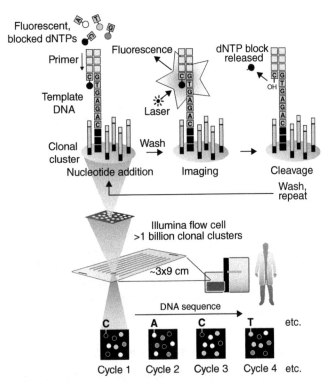

4. Reversible chain-termination sequencing (Illumina).

dNTP incorporation, these clusters are analogous to the bead-containing wells in the picotiter plates used in pyrosequencing. Illumina sequencing is initiated by adding to the flow cell a mixture containing all four dNTPs, each with a distinct fluorescent chemical block. Each DNA cluster emits one of four fluorescent signals, corresponding to each of the four possible dNTP incorporation events. Unincorporated dNTPs are removed and the entire surface of the flow cell is imaged using a camera; this serves to record which of the four fluorescent signals was detected at each cluster. The final step is to remove the chain-terminating blocks and wash away excess reagents, such that the next cycle can be initiated. The DNA sequence read length is determined by the number of cycles carried out.

As its chemistry is based on reversible terminators, Illumina sequencing does not suffer from the homopolymer problem; only a single dNTP can be incorporated at a time, regardless of the length of the homopolymer tract in a given stretch of template DNA. It does nevertheless have its limitations. Illumina read lengths are somewhat shorter than most other next-generation methods (initially 50 bp, currently upwards of 300 bp), which presents challenges when it comes to genome assembly (Chapter 3). What it loses in read length Illumina makes up for in terms of quality and throughput. Although a single run can take several days, the latest sequencers can generate up to four *billion* individual sequence reads and >500 billion bp of raw sequence data. It is this remarkable output that has made it possible to sequence a human genome in less than a week for ~US$1,000.

Single-molecule sequencing

The early years of next-generation tech development were a free-for-all of creative solutions to hard problems. Biotech companies hired teams of molecular biologists, chemists, and engineers to work together (and against other companies) to figure out how to best tweak their methods in order to maximize sequence read length, read quality, and output (via parallelization). We have explored only the most prominent of the next-generation sequencing methods that emerged during the early to mid-2000s, but there are two main issues common to all such technologies: short read lengths and the need for DNA amplification. In order to overcome 'signal-to-noise' problems inherent in the various systems, thousands of identical DNA fragments are deposited in a localized area (e.g. a flow cell well) and their collective 'signals' detected (e.g. light or fluorescence emission, pH change). Amplification is time-consuming and also has the potential to introduce biases, such as over- or under-representation of certain regions of the genome of interest due to variations in the relative abundance of As, Cs, Gs, and Ts. This brings us to what are sometimes referred to as 'third-generation' technologies, which are so sensitive as to be able to sequence DNA with single-molecule resolution. The technological challenges and potential pay-offs are significant.

The 'single-molecule real-time' sequencing device produced by Pacific Biosciences (PacBio) uses a flow cell covered with tiny wells a mere 50 nanometres across. Each well has a single, specially engineered DNA polymerase attached to its transparent bottom. Single-molecule sequencing is possible because the polymerase is stationary: the instrument's camera is able to focus in on what is happening at the bottom of each well, recording the fluorescence each time the polymerase incorporates a dNTP into the DNA chain (each dNTP has its own fluorophorescent tag). First-generation PacBio instruments are very large pieces of equipment (weighing ~1,800 pounds), but newer models are one-third the size and have improved chemistry and increased sequence throughput. Current flow cells have ~1 million wells each; average read lengths of 10–20 kilobase pairs (Kbp) are possible, with maximum reads exceeding 40 Kbp.

In 2014, Oxford Nanopore Technologies released the MinION, a device that like the PacBio can sequence single DNA molecules in real time. The MinION, however, can be carried in your pocket—it weights 90 grams, is $10 \times 3 \times 2$ centimetres in size, and plugs directly into a computer via a standard USB port. And unlike any other instrument available, the MinION determines the sequence of a DNA molecule by passing it through a protein nanopore (see Figure 5).

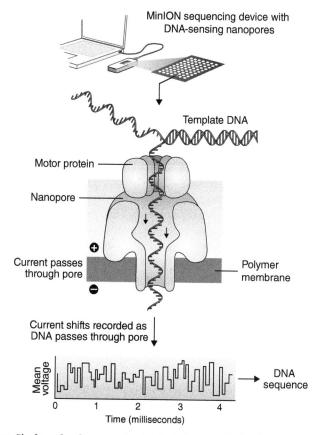

5. **Single-molecule sequencing (Oxford Nanopore Technologies).**

The procedure begins by applying an electric current to a synthetic membrane that has hundreds to several thousand such pores inserted in it. Specialized motor proteins deliver double-stranded DNA molecules to the nanopores, unzip the DNA, and ratchet them through each pore one base at a time. Incredibly, the sequence of each DNA molecule is inferred by recording the disruptions in current that occur at each nanopore as the nucleotide bases pass through it. Read lengths approaching 1 megabase pair (Mbp) have been reported, and the portable nature of the MinION has brought DNA sequencing to places it has never been. For example, in 2015 MinION sequencers were used in the field to track an outbreak of Ebola virus in Guinea. The instrument has also been used on the International Space Station. Total sequence output on the MinION was initially modest but is increasing steadily as the technology matures, and more and more nanopores are squeezed onto the flow cell. Oxford Nanopore Technologies also produces a bench-top version of the sequencer (the PromethION), which is capable of running up to forty-eight flow cells in parallel, each containing ~3,000 nanopores.

Despite their enormous potential, error rates for currently available single-molecule sequencers are higher than with Illumina and pyrosequencing, and the total number of reads generated per instrument run is typically lower. At the present time, single-molecule sequencing platforms have proven to be particularly effective when used in combination with more 'traditional' technologies such as Illumina. The idea is to combine the high quality and output of the latter with the longer read lengths generated by the former. As discussed in Chapter 3, read length becomes a critical factor when trying to assemble large, complex genomes.

Different next-generation sequencing platforms have different strengths and weaknesses. What is sufficient for decoding the genome of a virus may be woefully inadequate for a crop scientist studying a large plant genome. Factors such as sequence read length and quality, instrument run time and total data output, as well as differences in the amount of DNA template needed, influence one's choice of platform. Based on how rapidly technology has changed over the past decade, it is difficult to predict what DNA sequencing will look like more than a few years from now. But it is nevertheless safe to assume that it will soon cost as much to store DNA sequence information as it will to generate it.

Chapter 3
Making sense of genes and genomes

Sequencing a genome is like solving a puzzle. Some parts come together with ease, in the same way that a puzzle's edge pieces can be efficiently sorted and snapped together to frame the image. Other parts are akin to a uniform blue sky—they are difficult to assemble without systematic effort (and often leave one wondering whether essential pieces are missing). While DNA sequencing is faster and cheaper than ever before, genome assembly remains a significant challenge. In this chapter, we explore how laboratory and computational methods are used in combination to elucidate the true physical nature of DNA molecules inside living cells, and how genes are identified among the vast quantities of chemical letters that make up an organism's genome.

Shotgun sequencing

A genome cannot (at least not yet) be sequenced by starting at one end of the DNA and proceeding continuously to the other. Each string of nucleotides produced by a DNA sequencing instrument—each read—typically represents only a tiny snippet of the target genome. Furthermore, the imperfect nature of the methods used to distinguish between the As, Cs, Gs, and Ts of DNA means that each nucleotide position in the genome must be interrogated multiple times in order to produce a reliable sequence. For these reasons, the field of genomics has employed a procedure known as whole genome 'shotgun' sequencing. Named in reference to the random spread of pellets that emerge from a shotgun barrel, it was first used in 1981 to sequence the ~8,000 bp genome of the cauliflower mosaic virus. Although shotgun sequencing proved to be ideally suited to the task of sequencing small prokaryotic, mitochondrial, and chloroplast genomes, it was initially considered to be inadequate for tackling the larger, more complex genomes of eukaryotic nuclei. This has proven not to be the case, as illustrated in the race to sequence the human genome (see Chapter 4). The method has stood the test of time in terms of balancing efficiency and accuracy.

Shotgun sequencing begins by isolating genomic DNA, randomly shearing it, and purifying a population of uniformly sized fragments. Traditionally, the fragments are then inserted ('cloned') *en masse* into circular plasmid molecules to form a genomic 'library'. From this large collection of plasmids, clones are chosen at random and the ends of the inserted DNA fragments are sequenced. With next-generation sequencing, the cloning step is skipped; sheared DNA fragments are randomly sequenced in a massively parallel fashion using one of the technologies described in Chapter 2. Regardless of which method is used, each sequence read represents a random snapshot of a small portion of the genome; if enough reads are obtained and compared to one another, the sequence of the original genome can be 'assembled' in the computer (see Figure 6).

The number of sequences one needs to generate depends on the size of the genome to be sequenced. It can be calculated ahead of time using the simple formula $N \times L/G$, where N is the number of DNA sequence reads, L is the average read length, and G is the length of the target genome. From this one can determine how much sequencing must be carried out in order to achieve the desired depth of genome 'coverage'. Here coverage refers to the

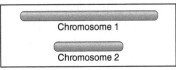

Genome to be sequenced

Chromosome 1

Chromosome 2

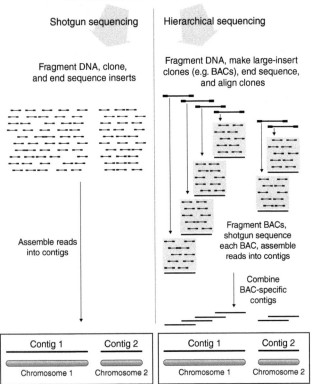

Shotgun sequencing

Fragment DNA, clone, and end sequence inserts

Assemble reads into contigs

Contig 1   Contig 2

Chromosome 1   Chromosome 2

Hierarchical sequencing

Fragment DNA, make large-insert clones (e.g. BACs), end sequence, and align clones

Fragment BACs, shotgun sequence each BAC, assemble reads into contigs

Combine BAC-specific contigs

Contig 1   Contig 2

Chromosome 1   Chromosome 2

6. 'Shotgun' versus 'hierarchical' approaches to genome sequencing and assembly.

average number of times that a particular nucleotide position in the genome is present in a collection of randomly obtained sequences. With traditional Sanger sequencing, approximately eight-fold coverage is sufficient to ensure that, simply by chance, ~97 per cent of the genome will be covered. Because next-generation sequencing techniques have higher error rates than Sanger's method, sequence depth must be higher if the same level of quality is desired (with Illumina, for example, it is necessary to obtain thirty- to fifty-fold coverage for the purposes of assembling a genome from scratch).

Genome assembly: problems and solutions

With sufficient coverage, the genomes of viruses and many prokaryotes can be accurately and efficiently assembled using raw data generated by most modern sequencing instruments. However, problems arise when repetitive DNA is encountered (i.e. when specific DNA sequences exist in more than one part of the genome). Most genomes contain repeats of some kind but the total amount varies considerably within and between prokaryotes and eukaryotes, as does the size of the repeat units. Nuclear genomes are notoriously repetitive; >50 per cent of the human genome is repeat sequences of various kinds (see Chapter 4) and some plant nuclear genomes have ~90 per cent repetitive sequences.

Why does repetitive DNA matter? Genome assembly algorithms detect overlaps between raw sequence reads and combine them to produce 'contigs', or contiguous sets of overlapping DNA fragments. A contig is a consensus of two or more DNA sequence reads. In principle, if enough reads are obtained, even an extremely large nuclear genome can be accurately assembled if it contains few or no repeats; individual contigs would correspond to whole chromosomes. However, the presence of repeats means that algorithms have difficulty determining from which region of the genome a particular repeat sequence is derived. If the average length of the sequence reads is shorter than the length of the repeats in the genome, the generation of large, accurate contigs is compromised. The more repeats, the bigger the problem.

In reality, most nuclear genomes are a mixture of unique sequences and short and long repeats of various kinds, and researchers have developed ways of dealing with the issue. One approach is to obtain sequences from both ends of the randomly selected DNA fragments (see Figure 6), and to have the algorithm keep track of these so-called 'paired ends' and the distance between them (which is known because the genomic DNA was size-selected prior to library construction and sequencing). This improves the chances of being able to anchor the sequence data from each DNA fragment onto the genome. It is also common to perform multiple sequencing runs using DNA that has been sheared to different sizes (e.g. 2, 10, and 50 Kbp); this makes paired-end data even more effective in overcoming the problems associated with repetitive DNA.

Even when these and other measures are taken, nuclear genomes are only rarely considered truly 'complete' in the sense that the sequence of each and every chromosome has been unambiguously determined from one end to the other. The term 'draft' is often used in reference to a genome that has been sequenced and assembled to a level of quality that allows most of the genes to be confidently identified and studied. Gaps remain but the end product is sufficient for most research applications.

The human genome sequence is a draft sequence. After shotgun sequencing the fruit fly genome as a proof of principle in 2000, the American scientist-entrepreneur Craig Venter successfully led a privately funded team of researchers in sequencing the human genome using the same approach. Their methodology differed from the hierarchical, 'top-down' strategy employed by the publicly funded Human Genome Project. In this case, large DNA fragments 50–200 Kbp in length were cloned into special vectors called 'bacterial artificial chromosomes' (BACs). By sequencing snippets of the ends of thousands of such 'large-insert' clones, a low-resolution physical map of the chromosomes was inferred prior to high-throughput sequencing. Individual BAC clones were then selected and shotgun sequenced on a small scale as necessary in order to fill in the remaining gaps along each chromosome (see Figure 6).

This process sounds laborious and it was! But it is important to recognize that the publicly funded project began at a time when Sanger sequencing was still arduous and expensive; it was thus critical to minimize the time and money spent generating raw sequence data. The BAC-based approach also allowed different laboratories around the world to work in parallel without duplicating their efforts (e.g. by assigning different BACs and chromosomes to different teams). This is not possible with a purely random shotgun-sequencing approach. Today BAC clones are only rarely used in standard genome projects (although they are an important aspect of the 'metagenomic' sequencing of DNA taken directly from the environment; see Chapter 6). Long-read sequencing technologies such as those developed by PacBio and Oxford Nanopore Technologies (Chapter 2) are efficient and increasingly effective tools with which to address the repetitive DNA problem, and 'standard' next-generation methods such as Illumina are being used in creative new ways to improve the quality of genome assemblies.

One approach, developed by a company called 10x Genomics, involves using a microfluidics device to partition large fragments of the genome into tiny 'gel beads'. The DNA contained in each bead is amplified, tagged with a unique 'barcode', and sequenced on an Illumina instrument. The production of barcoded sequence reads greatly enhances the genome assembly process by providing long-range chromosomal information: reads with the same barcode are derived from the same gel bead and the same physical piece of DNA. They can thus be confidently mapped to the same region of the genome. Remarkably, this approach has recently been used to produce seven high-quality human genome assemblies at low cost and starting with as little as ~1 nanogram of DNA.

Gene finding with transcriptomics

In the 1980s and 1990s, when genome sequencing was still in its infancy, techniques were developed towards the goal of speeding up the process of gene discovery in humans. The idea was to target only those regions of the genome that are transcribed into RNA, thereby focusing directly on protein-coding regions and avoiding the problem of having to generate (and sift through) massive amounts of non-coding genomic DNA in search of genes. The technique involves using RNA as a template for the production of so-called complementary DNA (cDNA). This step is carried out by reverse transcriptase, a DNA-synthesizing enzyme that exists naturally in viruses with RNA genomes. Reverse transcription of RNA into DNA is both necessary and useful: there is no easy way to sequence mRNAs directly (Woese's RNA cataloguing technique was only applicable to particularly abundant RNAs and was abandoned when Sanger DNA sequencing came along), and DNA is generally much more stable to work with in the lab than RNA.

Once synthesized, cDNA molecules can be cloned into plasmids, propagated in *E. coli*, and easily sequenced. The advantage of the approach is that the cDNAs are devoid of introns, which are conveniently spliced out of the mRNAs in the cell prior to reverse transcription in the test tube. The 'bang for the buck' is thus huge—the cDNAs contain only the protein-coding exons, regardless of the size and abundance of introns in the genes themselves. Individual cDNA sequencing reads were referred to as expressed sequence tags (ESTs), and the approach became known as EST sequencing.

Although initially applied in biomedical research contexts, EST/cDNA sequencing has become a go-to technique for researchers working all across the life sciences. It is particularly useful in the study of gene expression in multicellular organisms. RNA samples can be extracted from different tissues, converted to cDNAs for sequencing, and the relative abundance of cDNAs (each derived from an mRNA transcript) in the different datasets can be compared. Such an approach can yield important insights into the genes and proteins that are active at different times in different parts of the organism. EST sequencing has also become a popular and affordable alternative to genome sequencing in diverse areas of biology. The sequencing of just a few thousand clones randomly selected from a traditional cDNA library is enough to provide a sneak-peak at the gene repertoire of a newly discovered organism without the expense and complexities of tackling a genome.

Up until the mid-2000s, ESTs were sequenced primarily using traditional Sanger sequencing, but it too has evolved in response to new DNA analysis technologies. The approach has now been replaced by what is called RNA-seq (RNA sequencing), which is a misnomer in that, like the EST approach, it relies on the conversion of RNA to cDNA prior to sequencing. The main difference between traditional EST sequencing and RNA-seq approaches is that the latter employs next-generation sequencing technologies and thus has the ability to generate much more data in a given experiment. With a technology such as Illumina, it is possible to detect mRNA transcripts corresponding to 80–90 per cent of the genes present in an organism's genome, even those that are expressed at very low levels at the time that RNA was extracted. It is also possible to quantify differences in the level of expression of thousands of genes simultaneously, for example, in the different tissues of a multicellular organism or between cell cultures grown under different environmental conditions. The term *transcriptomics* refers to the study of the transcriptome (i.e. the complete set of RNA molecules present in a cell or population of cells).

RNA-seq has become an indispensable tool that goes hand in hand with genome sequencing, greatly accelerating the process of

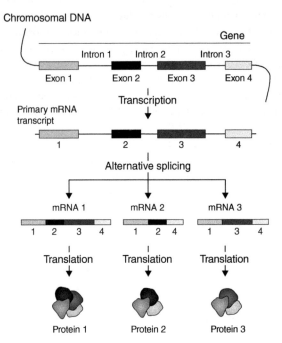

Genomics

7. **Alternative splicing.**

finding genes in genomes of all shapes and sizes. Predicting
intron-exon boundaries has turned out to be more complicated
than simply aligning cDNA and genomic DNA sequences. This is
in large part due to the phenomenon of alternative splicing, which
is when the same gene gives rise to more than one type of mRNA
transcript (see Figure 7). Under different conditions or in different
tissues, a single gene can yield dozens of functionally distinct
proteins by 'mixing and matching' exons. With RNA-seq, we now
know that alternative splicing is a feature of all eukaryotic cells,
from single-celled microbes to humans.

RNA-seq has also proven to be a highly effective means with
which to identify small RNA genes such as those encoding the sea

of microRNAs that are emerging as important regulators of the expression of protein-coding genes. Because small RNA gene sequences often differ greatly from species to species, they can be difficult to find using standard sequence similarity searches. In this case, researchers begin by size-selecting a fraction of RNA that is much smaller than the protein-coding mRNAs. The small-RNA fraction is then converted into cDNA as per usual and deeply sequenced using a suitable next-generation technology. By considering the relative abundance of the small RNA transcripts and the locations to which they map on the genome assembly, it is also possible to link small RNAs to the specific protein-coding genes whose expression they regulate.

Where are the genes and what do they do?

At the heart of modern genomics is a computer program called BLAST (basic local alignment search tool). Here molecular sequences are used to query a database containing gene sequences from other organisms in order to identify regions of similarity. The BLAST algorithm searches for regions of local alignment between the query and one or more sequences in the database. A list of 'hits' is produced, each assigned a measure of statistical significance called an 'expect' value (E-value). Simply put, the E-value describes the number of hits one would expect to see simply by chance when searching a database of a particular size. The smaller the E-value for a given BLAST hit, the more 'significant' the match is and the more confident the researcher can be that a real gene has been identified. BLAST searches can be performed using DNA sequences as queries (which is necessary when searching for non-coding RNA genes) or the amino acid sequences of proteins; the latter is more common for genome-scale analyses, in part because it is better able to detect sequence similarity over greater evolutionary distances.

Genes can also be found using so-called *ab initio* approaches. In this case, software is used to recognize the features of genes in the genome of interest such as the nucleotide composition of intron-exon boundaries. Prior to gene finding, the software is 'trained' on a curated set of known genes and their mRNA transcripts; once it has 'learned' what the organism's genes typically look like, it can be used to identify all of the stretches of DNA sequence in the genome that have these features, regardless of whether or not their sequences are similar to known genes in other organisms. The *ab initio* approach has its limitations, most notably the fact that at least some gene-sequence data must already been in hand (otherwise the computer algorithms cannot be trained). In practice, the process of gene prediction typically involves running several different types of gene-finding software and comparing the results. Stretches of sequence that are predicted to be genes on the basis of multiple lines of evidence are considered more likely to be *bona fide* genes, especially if there is clear evidence for transcription.

Having predicted gene sequences, the next step is to assign function. To a certain extent, this emerges naturally from the BLAST-based procedures used to identify the genes in the first place; if the sequence of a particular gene/protein from species X is similar to a protein of known function from species Y, then it is likely to have the same or similar function. BLAST results are typically linked to large databases containing many thousands of curated protein sequences, proteins for which function(s) have been confirmed by experimentation. Often functions have been assigned to particular sub-regions of a protein—protein domains. For example, the Pfam database contains ~17,000 distinct protein 'families' and serves as an important reference in genome annotation pipelines. Proteins often have multiple domains with distinct sub-functions. For example, transcription factors are proteins that play important roles in regulating gene expression; they typically have DNA-binding domains and protein-protein interaction domains, both of which can be detected using computer programs.

38

For a newly sequenced genome, the goal is to assign a putative function to as many of the genes as possible. The level of confidence associated with these assignments varies greatly. Sometimes it is possible to predict with a high degree of certainty that gene X encodes a particular DNA replication protein. Often it is not possible to be that specific. A protein may be predicted to be a sugar transporter (because it shares amino acid sequence similarity with transporters characterized in other organisms) but it may not be clear which exact type of sugar it transports. Or a protein may be predicted to bind to cell membranes (because it has putative membrane-spanning domains), but with no further functional insight being possible. And often a predicted protein has no obvious function whatsoever. This is a particularly common occurrence when annotating the genome of an organism that has no known close relatives. A well-annotated genome includes information on a wide range of different biological processes, including predicted metabolic capacities (which can be used to infer how the organism makes a living) and cellular features such as the capacity to make a flagellum with which to move about.

Where do the proteins function?

Another aspect of genome annotation is predicting where protein molecules carry out their functions. This is possible because proteins often contain specific amino acid sequences that serve as intracellular 'postage stamps'. For example, most (but not all) proteins that function in the mitochondrion are actually encoded in the nuclear genome and synthesized in the cytoplasm; such proteins contain a fifteen to thirty amino acid 'targeting peptide' on one end that serves to direct them to the organelle. Nucleus-encoded chloroplast proteins have similar targeting sequences. Algorithms have been trained to scan the amino acid sequences of proteins for the presence of these and other targeting elements, and can be used to predict which among a set of thousands of proteins are most likely to function in a particular organelle. Another example is the KDEL motif; this four amino acid sequence is the hallmark of proteins that reside within a particular region of the endomembrane system of eukaryotic cells. Although not perfect, researchers use these tools to assemble protein inventories corresponding to discrete regions of the cell. These predictions can then be used to formulate hypotheses for testing in the laboratory.

One experimental method used to confirm subcellular location is a rapidly maturing 'omics' technology called proteomics. In this case, a protein-cutting enzyme such as trypsin is used to digest a sample of proteins that have been purified from an organelle such as the mitochondrion. The fragmented mixture is then run through an instrument called a mass spectrometer. Because of the specificity of the trypsin enzyme, which cuts proteins only at particular amino acids, the protein fragments are of a defined size, and can thus be detected by the instrument. By comparing the experimental output with proteins predicted from genome sequences, one can verify which proteins are indeed localized to the organelle of interest. More generally, proteomics is used hand in hand with genomics to confirm that computer-based gene predictions are accurate—false positives will not produce real proteins.

Genome re-sequencing

DNA sequence comparisons between two or more members of the same species are a rich source of information for the study of molecular and evolutionary biology. But even with the best tools and brightest minds, the challenge of assembling a genome *de novo* (i.e. from scratch) can be prohibitive. Fortunately, there are situations in which it is unnecessary. Genome 're-sequencing' involves generating raw sequence data from the individual or strain of interest and 'mapping' the reads onto a pre-existing assembly. Re-sequencing can be done in a targeted fashion (e.g. by focusing only on the exons) or it can be applied to the whole genome. Generally speaking, the higher the quality of the reference genome assembly, the more useful a re-sequencing strategy will be. Although often overlooked, our ability to 'sequence' a human genome in a week for a thousand dollars is only possible because of the genome mapping efforts of the original publicly funded Human Genome Project, which was itself built upon the work of human geneticists decades prior.

A targeted approach called 'whole-exome sequencing' is used extensively in human disease research. This can involve use of a 'microarray', a small piece of glass to which single-stranded DNA molecules corresponding to a defined set of exons are permanently bound. Arrays containing >100,000 human exons are commercially available and used for diagnostic purposes. DNA samples from patients are flooded over the microarray, material that has not 'hybridized' (i.e. base paired) to the DNA on the array is washed off, and the DNA that remains is collected and next-generation sequenced. In essence, whole-exome sequencing allows researchers to efficiently screen the protein-coding regions of the genome for disease-causing mutations at a fraction of the cost of whole-genome sequencing.

Whole-genome re-sequencing is fast becoming the norm in the field of human comparative genomics. It is sufficient for the accurate detection of single-nucleotide polymorphisms or SNPs (i.e. variations in a single nucleotide position in the genome in members of a population). To a lesser extent it can also identify short insertions and deletions in non-repetitive DNA, as well as reveal copy number variations (CNVs) between target and reference genomes. CNVs are increasingly recognized as being important in human biology and disease; in genomic data, they manifest themselves as differences in the depth of sequence coverage obtained for a specific gene or stretch of DNA. Genomic regions showing a large increase or decrease in the number of individual reads mapping to the reference genome

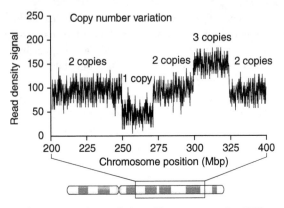

8. Detecting copy number variation with next-generation DNA sequencing.

are indicative of changes in copy number in the organism of interest (see Figure 8).

Genome re-sequencing has its limitations. For example, expansions and contractions in repetitive sequences are largely invisible with this methodology; if individual DNA sequence read lengths are significantly shorter than the length of an individual repeat unit, read-mapping data alone will not provide meaningful information on changes in repeat number from genome to genome. The issue is confounded by the fact that the quality of the reference genome may be compromised in such regions due to the presence of repeats.

Another type of structural variation that can be difficult to detect with a generic re-sequencing approach is inversions. These can be as small as a few tens of nucleotides or over a million bp in size; how reliably inversions of different lengths can be detected in genomic data depends on factors such as sequence read length and the computational tools used to map the reads onto the genome. Because inversions can disrupt gene sequences and/or influence the expression of nearby genes, they are of considerable interest to researchers studying the genetic basis of disease. In humans, the largest known inversion is a ~4.5 million base-pair stretch of DNA on chromosome 8 that is flipped (relative to the reference genome) in an estimated 12–60 per cent of the population, depending on ethnicity. Although the underlying molecular mechanisms are unclear, this inversion is associated with rheumatoid arthritis and the autoimmune disease lupus.

Another well-studied example is the so-called Iceland inversion, a ~900 Kbp region of human chromosome 17 that is inverted in ~20 per cent of Europeans. A 2005 study showed that Icelandic women with the chromosome 17 inversion have increased rates of DNA recombination and, for reasons unknown, give birth to more children. In contrast, the 'standard' (i.e. non-inverted) form of chromosome 17 is associated with several neurodegenerative disorders, including Alzheimer's and Parkinson's disease. A common feature of these and other inversions is the presence of repetitive sequences at their breakpoints, which are difficult to map onto the reference genome. An active area of bioinformatics research is the development of better algorithms for detecting both large and small genomic inversions from next-generation sequence data; such algorithms continue to evolve along with sequencing approaches aimed at collecting long-range information such as the 10x Genomics approach describe earlier. As we shall see in Chapter 4, the human genome sequence is still very much a 'work in progress'.

Chapter 4
The human genome in biology and medicine

Two projects, one vision

The initial phase of human genome sequencing is often referred to as 'the' Human Genome Project. But there were of course two different projects, one publicly funded, the other supported by a private company. There is also no simple answer to the question of whose genomes were sequenced. The public project was based on DNA samples extracted from white blood cells donated by a number of anonymous men and women. The bulk of the sequence data is reportedly derived from a single male donor in Buffalo, New York. In the case of the competing private project, carried out by Craig Venter's Celera Genomics, DNA was sequenced from five individuals (including Venter himself).

Recall that the two projects differed significantly in their over arching strategies. The public project relied heavily on the labour-intensive production of a so-called physical map, a higher-order representation of the structure of each of the chromosomes upon which smaller segments of assembled sequences could eventually be mapped. It took thirteen years (1990–2003) and an estimated US$3 billion to complete. In contrast, Venter's US$300 million project, started in 1998 with cutting-edge Sanger sequencers, was a shotgun affair—the effort was focused on generating individual sequence reads and assembling them into contigs and larger scaffolds. (Venter's assembly is purported to have relied on preliminary data released by the public project; this was a point of contention since it was, technically speaking, not a true test of a shotgun-only approach.) The two assemblies differed in terms of the number of gaps they contained, their physical structures, and the number of genes they predicted. All things considered, however, they were very similar; both genomes were deemed ~90 per cent 'complete' at the time of publication in 2001.

Although considered necessary in order to minimize legal and ethical concerns, the fact that both projects used DNA samples from more than one person resulted in the production of composite genome sequences. In 2007, Craig Venter's team published the first genome of a single individual (that of Craig Venter). The genome sequence of James Watson, the American co-discoverer of the structure of DNA, was published shortly thereafter and was noteworthy in being the first to be done using a next-generation technology (pyrosequencing; see Chapter 2). The Venter and Watson genomes thus marked the beginning of the era of personal genomics, which is proceeding with ever-increasing speed and sophistication. The first two human genome projects produced 'the' genome—a genome sequenced for all of humanity. But there is in fact no such thing as 'the' human genome. This reality underlies the concept of personalized (or precision) medicine.

Human genome structure

The ~3.2 billion bp of nuclear DNA in humans is distributed across twenty-three pairs of chromosomes (see Figure 9). Except for eggs and sperm, humans cells are diploid, meaning that they have two copies of each chromosome; there is thus ~6.4 billion bp of DNA present in each cell nucleus. The so-called autosomes are numbered 1 to 22 and range from ~250 Mbp (chromosome 1) to ~50 Mbp (chromosome 22) in size. Although very different from

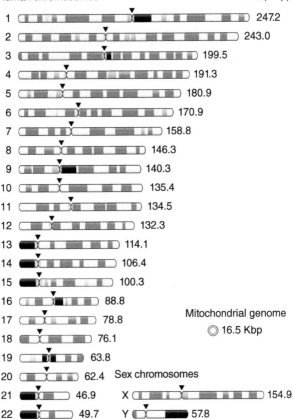

9. **Structure of the human genome and its chromosomes.** Humans contain twenty-two autosomes and two different types of sex chromosomes, X and Y. The arrowheads highlight the chromosome centromeres. Grey-shaded regions correspond to distinct chromosomal regions that are visible upon cytogenetic staining. Black regions correspond to the largest regions of the human genome that have not yet been sequenced and/or contain unresolved sequenced regions, such as the ribosomal RNA-encoding repeats. The human mitochondrial genome is also shown (not to scale).

one another, the two sex chromosomes, X and Y, are considered a pair. Research has shown that they diverged from what was originally an identical pair of autosomes more than 180 million years ago, early in the evolution of mammals. Females have two copies of the X chromosome per cell (one present in the form of a condensed structure called a Barr body) while males have one X and one Y.

The X chromosome is 153 Mbp in size and has ~2,000 genes. The Y chromosome is less than half the size (59 Mbp) and is something of a genomic wasteland: as of 2003, its gene count stood at seventy-eight. Of all human chromosomes, Y has proven the most difficult to sequence. It is full of repetitive regions and even with modern technologies, more than 50 per cent of it remains un-sequenced (see Figure 9). With careful scrutiny a handful of additional genes have recently been identified, a trend that is likely to continue as more of the Y chromosome is characterized. Despite its degenerate nature, the Y chromosome is clearly the sex-determining chromosome. One particular gene, SRY (sex-determining region Y), encodes a protein that is involved in the development of testes, and various other Y-localized genes play a role in the making of sperm.

Human chromosomes 13, 14, 15, 21, and 22 each have a large region of unresolved sequence at one of their ends. These are the locations of the ribosomal RNA-encoding genes, which are present in tandem-repeated segments whose precise structures have not yet been puzzled out. Additional large, un-sequenced portions of several chromosomes reside near the centromeres (see Figures 2 and 9), which are genomic regions that are important for the partitioning of chromosomes to daughter cells during cell division. Various smaller un-sequenced parts of the genome are sprinkled across the twenty-two autosomes and sex chromosomes. In general, such regions are characterized by repetitive DNA and contain few or no genes. There are even small bits of genome that have been sequenced but for various reasons have not been mapped to a specific chromosome. In 2015, researchers capitalized on the long reads produced by the single-molecule PacBio sequencer to resolve the structure of >50 per cent of the gaps that remained in the reference genome (the sequence derived from the publicly funded project is the reference to which others are usually compared). More than fifteen years on, however, there are still numerous regions of the human genome that remain un-ordered and un-sequenced.

The human mitochondrial genome

It is easy to forget that each of us has *two* genomes. We have been discussing the DNA that resides in the cell nucleus, but our mitochondria also contain DNA. Mitochondria are important double-membrane-bound compartments whose best-known role is the production of ATP, the energy currency of the cell. ATP is used to fuel chemical reactions in all of our different cell types, and indeed, in all cellular life forms. Mitochondria evolved from once-free-living bacteria by endosymbiosis (see Chapter 5). The mitochondrial genes of all eukaryotic organisms, including humans, are very clearly most similar to those of a particular group of bacteria called α-proteobacteria.

The human mitochondrial genome is a circular molecule 16,569 bp in size (see Figure 9); it contains only thirteen protein-coding genes, two rRNA genes, and twenty-two genes for transfer RNAs that bring amino acids to the ribosome during protein synthesis. Most of the genes for the 1,000+ proteins needed to maintain a functional mitochondrion are located in the nuclear genome. These proteins are translated in the cytoplasm of the cell and targeted to the organelle after they have been synthesized. From a biomedical perspective, an increasing number of diseases have been shown to be associated with mutations in mitochondrial genes and nuclear genes for mitochondrial proteins. The inheritance patterns of such mutations differ depending on the location of the gene. This is because in humans and many other animals, offspring receive their mitochondria—and mtDNA—exclusively from their mothers. This so-called 'maternal inheritance' means that males do not pass on mtDNA mutations to their children. In contrast, mutations in nuclear genes for mitochondrial proteins can be inherited from either parent in a standard Mendelian fashion. As we shall see in Chapter 5, the specific properties of mtDNA have proven useful in diverse scientific areas, including forensics, genealogy, and the study of human migration patterns.

How many genes?

The question of how many genes reside in our nuclear genome is the subject of ongoing debate. Early predictions of between 40,000 and 100,000 protein-coding genes proved to be significant over estimates. From their new-and-improved sequences published in 2003, the private (Venter) and public human genome projects predicted the existence of ~26,000 and 30,000 genes, respectively. However, as of this writing there are just over 20,000 human protein-coding genes annotated in public databases, and a 2014 proteomics-based study suggested that the true number of genes might be as low as ~19,000. These estimates do not include most of the ~13,000 predicted pseudogenes in the genome (i.e. genes that have been inactivated due to mutations). Many pseudogenes are the product of gene duplication; in some of the largest multi-gene families more than half of the genes are pseudogenes. Nor do these numbers include the myriad newly discovered genes for non-coding RNAs, many of which were initially predicted to be protein coding.

New protein-coding genes will no doubt be discovered as the human reference genome becomes more refined and a consensus emerges as to the most accurate gene-finding algorithms. Other genes will be crossed off the list as false positives. But regardless of the precise number, our gene count is on par with that of many other 'simpler' organisms such as fruit flies (~14,000 genes), nematode worms (~20,000), and the model plant *Arabidopsis* (~26,000). What are we to make of this? Despite having only ~20,000 genes, research suggests that human beings are comprised of at least 100,000 distinct proteins. This means that, on average, each of our genes is capable of making multiple, functionally distinct proteins by alternative splicing. These different proteins are often expressed in different tissues, the result of tissue-specific splicing factors that direct the splicing machinery to produce different mRNAs by mixing together different exons (see Figure 7). Alternative splicing thus helps to explain how 'complexity' can vary considerably between organisms with similar numbers of genes.

Jumping genes and 'junk' DNA

One of the truly remarkable features of the human genome is just how little of it is obviously 'useful'. Only ~1 per cent of the genome codes for protein; almost a quarter of the genome is comprised of introns (the sequences that interrupt the protein-coding exons and are spliced out of the mRNA prior to translation) (see Figure 10). And almost half of the human genome is made up of transposable elements. These genetic elements are sometimes referred to as 'jumping genes' and for good reason: they are DNA

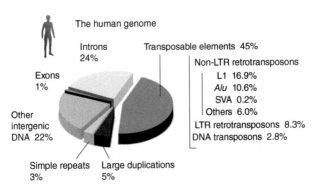

10. **Composition of the human genome. The various classes of transposable (mobile) genetic elements are highlighted.**

sequences that can change position. So-called retrotransposons are the most abundant class of transposable element in the human genome. Retrotransposons duplicate themselves via a 'copy-and-paste' mechanism in which they are transcribed into RNA, reverse transcribed into DNA, and then inserted into a new location in the genome.

Many of the retrotransposons in our genome are dead or dying; their mutated sequences are no longer capable of movement and are disappearing into the genomic background. Others are active and spreading like weeds. Among the most prominent type of retrotransposon is *Alu*, a short element present in more than a million copies scattered across our chromosomes (this single 300 bp transposable element makes up almost 11 per cent of the entire genome). Comparative genomics has revealed that *Alu* arose from the fusion of two pieces of a small RNA gene ~65 million years ago; it has been propagating itself within primate genomes ever since. Other retrotransposons, such as L1, clearly evolved from retroviruses, special viruses that spend part of their 'lives' inserted into the genome of their host. Retroviral sequences make up ~8 per cent of the human genome.

Collectively, transposable elements can be considered 'selfish' genetic elements: they spread simply because they can and, assuming they do not over-burden their host, they will continue to proliferate. An increasing number of *de novo* retrotransposon insertions have been shown to be responsible for human diseases. For example, *Alu* insertions have been linked to certain types of haemophilia, colorectal cancer, breast cancer, leukaemia, and cystic fibrosis. L1 retrotransposition events have been shown to cause colon cancer and X-linked Duchenne muscular dystrophy. In some cases, the problem stems from the transposable element having inserted itself directly into a protein-coding gene, disrupting its ability to produce a functional protein. In others, the element has inserted into non-coding 'regulatory' DNA, thereby modifying the expression of critical genes nearby. While there is still much to learn about transposable elements, it is clear that their activities have impacted—and continue to impact—the biology of humans and other primates.

In 2003, the US National Human Genome Institute launched the Encyclopedia of DNA Elements (ENCODE) project. On paper, the goal was simple: identify all of the functional elements in the human genome. In practice, this involved using a battery of experimental and bioinformatic approaches to try and better understand the non-protein-coding portions of the genome. Having undergone several iterations, the ENCODE project has led to the discovery of many thousands of putative small RNA genes hiding in what was once considered non-coding DNA. Most have yet to be verified experimentally, but some of these small RNAs have been shown to regulate the expression of protein-coding genes.

The ENCODE project's broadest and most controversial claim is that 80 per cent or more of the human genome has a biochemical function. The claim is based on the observation that ~80 per cent of the genome (i) shows evidence of being transcribed into RNA; (ii) is bound to proteins; or (iii) is contained within specially modified chromatin. The results were portrayed (mainly in the media) as the death of 'junk' DNA, a loaded term that has long been used in reference to the large fraction of the human genome (and those of many other eukaryotes) with no discernable function. Much of what is generally considered junk DNA is clearly derived from the selfish spread of transposable elements; the 45 per cent described earlier (see Figure 10) only includes those that can be detected by sequence similarity (the true fraction is probably much higher). Research has also shown that non-functional pseudogenes can be expressed. The ENCODE claims have thus ruffled the feathers of more than a few evolutionarily minded genomicists who object to (among other things) the project's use of the term 'function'. Although nebulous, the concept of biological function is central to the definition of a gene, which itself continues to evolve. The debate is sure to continue as scientists further unravel the complexities of the non-coding portions of our genome.

The HapMap to personalized medicine

From its inception, a principal goal of the Human Genome Project was to transform biomedical research and, ultimately, the practice of medicine itself. To that end, the International HapMap Project was built off of the human reference genome as a platform for determining genetic variants linked to health and disease. This was achieved by generating a so-called 'haplotype map'. A haplotype is a set of specific DNA sequences that are inherited as a unit. By surveying genetic diversity across different human populations, common patterns of DNA sequence variation can be discerned. The information can be used to improve the power of genome-wide association studies (GWAS), which are used to identify associations between SNPs and human traits—everything from height and hair colour to high blood pressure and the consistency of earwax. SNPs are defined as changes in a single nucleotide position that are present in >1 per cent of the population. Many SNPs result in changes in a protein sequence; most do not.

The HapMap Project was initiated in 2003, prior to the advent of next-generation sequencing. It was thus modest in terms of the number of people studied (initially ~270 individuals) and the amount of data collected (hybridization-based genotyping arrays were used, as opposed to actual DNA sequencing). With technological advancements, the density and usefulness of the map has improved dramatically. Hundreds of thousands of SNPs have been identified, more than a thousand of which have been linked to specific disease phenotypes. GWA studies have given biomedical researchers a better foothold on the genetic foundations of diseases such as sickle-cell anaemia, cystic fibrosis, and certain heritable forms of cancer. But GWA studies also have their limitations. They are helpful in identifying *associations between* DNA sequence variation and diseases but they cannot, in and of themselves, specify *cause*. The success of GWAS depends on properly designed sampling regimes that include large numbers of individuals in both disease-affected and control groups. More generally, GWA studies have been criticized for placing undue emphasis on the genetic aspects of health and disease, and in so doing down-weighting the importance of factors such as diet and environment.

Genomes by the thousand

As strange as it sounds, it is no longer possible to determine how many human genomes have been sequenced. At present the strategy of choice is whole-genome re-sequencing (Chapter 3) whereby next-generation sequence data are mapped onto a reference genome. The results have been breathtaking. The recently concluded (and aptly named) 1000 Genomes Project Consortium catalogued ~85 million SNPs, 3.6 million short insertions/deletions, and 60,000 larger structural variants in a global sampling of human genetic diversity. These data are catalysing research in expected and unexpected ways. Beyond providing a rich source of data for GWA-type studies focused on disease, scientists are also using the 1000 Genomes Project data to learn about our basic biology, something that proved surprisingly difficult when only a pair of genomes was available. For example, a recent GWAS taking advantage of the 1000 Genomes Project data identified ten genes associated with kidney development and function, genes that had previously not been linked to this critical aspect of human physiology.

In 2016, Craig Venter's team reported the sequencing of 10,545 human genomes. Beyond the impressively low cost (US$1,000–2,000 per genome) and high quality (30–40× coverage), the study was significant in hinting at the depths of human genome diversity yet to be discovered. More than

150 million genetic variants were identified in both coding and non-coding regions of the genome; each sequenced genome had on average ~8,600 novel variants. Furthermore, each new genome was found to contain 0.7 Mbp of sequence that is not contained in the reference genome. This underscores the need for methods development in the area of structure variation detection in personal genome data. Overall, however, the authors concluded that 'the data generated by deep genome sequencing is of the quality necessary for clinical use'.

As of this writing, the website ticker for the UK 100,000 Genomes Project tells me that they have sequenced 39,540 human genomes (I'll check back in an hour in case the number has increased). This ambitious government-funded initiative is sequencing the genomes of National Health Service patients, with an emphasis on rare and infectious diseases as well as certain types of cancer. There is widespread agreement that of all human afflictions, it is the diagnosis and treatment of cancer that has the most to benefit from genome sequencing.

Cancer genomics

It is common to think of cancer as a single disease, but it is in fact a constellation of many diseases characterized by uncontrolled cell growth, tissue invasion, and metastasis (spread). Hundreds of different cancers and tumour types have been defined, the vast majority of which clearly involve DNA modifications. Pre-existing inherited mutations can predispose individuals to developing cancer, but most cancers stem from the accumulation of mutations and other genomic alterations during one's lifetime. For these reasons, genomics research is having a profound impact on our understanding of the genetic, biochemical, and cell biological underpinnings of cancer, and how it can be detected and treated.

There are two basic types of genes that contribute to cancer development and progression. Tumour-suppressor genes are those that, when mutated or otherwise inactivated, give rise to uninhibited cell growth. Tumour-suppressor genes, such as the gene that encodes a protein called p53, often encode transcription factors (i.e. proteins whose job is to regulate gene expression). Transcription factors are molecular 'hubs': mutations that turn them on or off have far-reaching implications by modifying the expression of other genes, which can in turn trigger further cellular perturbations. Tumour protein p53 has been described as 'the guardian of the genome'; almost 50 per cent of known cancers involve a p53 gene mutation (or some indirect disruption of normal p53 protein function). In contrast, proto-oncogenes are genes whose protein products lead to cancer progression only when over-activated; they typically belong to biochemical signalling pathways involved in cell division. Like tumour-suppressor genes, proto-oncogenes are 'normal' genes. They become oncogenes only if they are mutated in a way that leads to unregulated cellular proliferation.

Cancer-associated mutations need not occur within gene sequences. Because non-coding regions can be important in regulating the expression of surrounding genes, mutations, insertions, or deletions of such 'regulatory' sequences can contribute to the development of cancer. Cancer also involves genomic changes that influence how the cell interacts with its surroundings. This includes evasion of the immune system, enhancing blood flow to the tumour (which further aids growth), and increasing the probability of tissue invasion and metastasis. Whole genome-scale analyses of diverse tumour types have begun to shed light on how these events are triggered.

A landmark investigation was the recently completed Cancer Genome Atlas. Over an eight-year period starting in 2003, the project profiled the genomes of ~10,000 tumours, laying important groundwork for future developments in the field. Deep insight into the origins of cancer has subsequently emerged from the *de novo* sequencing of many hundreds of tumour genomes. It is now clear that there are ~140 human genes that, when mutated, can promote tumour development. Any given tumour has between two and eight of these so-called 'driver gene' mutations—all of which impact cell growth/survival or genome maintenance in some way—as well as dozens to thousands of additional single-nucleotide changes ('passenger' mutations) and small insertions/deletions.

More radical genomic alterations have also been shown to play important roles in cancer. A remarkable example is the 'HeLa' genome. HeLa was the first human cell line to be permanently established in culture; the name refers to Henrietta Lacks, the American patient from whom the original cells were derived (without her consent) and who died from cervical cancer in 1951. Because of their immortality, HeLa cells have for many decades been a mainstay of molecular- and cell-biology research.

Relative to the reference genome, the HeLa genome is characterized by ~1.7 million single nucleotide variants, ~750 large deletions, and a plethora of chromosome-scale rearrangements and duplications. Many of the alterations seen in HeLa, in particular, those involving chromosome 11, are also seen in other cervical cancers, as well as breast, nasopharynx, and head and neck cancers. HeLa cells are abnormal in that three or four copies of some of the chromosomes are present in each cell (between seventy-six and eighty chromosomes per cell in total).

Not surprisingly, transcriptomics has also shown that patterns of gene expression in HeLa cells differ greatly from the cells of normal tissues, a fact that can complicate the interpretation of research results obtained using HeLa as a model cell line. It is not clear how many of these genomic alterations stem from the original tumour or have taken place during the many years HeLa cells have been propagated in labs around the world. The extent to which chromosome rearrangements and duplications play a role in the development and progression of cancer (as opposed to being a consequence of it) is in many cases unclear.

Regardless, genome sequencing has confirmed what clinicians have long suspected: tumours evolve. With the latest technologies researchers have been able to 'replay the tape' of cancer from normal tissue to primary tumour through to metastasis. This has led to unexpected conclusions, including the fact that metastatic cancers can arise from multiple distinct cancer cell lineages within a primary tumour. This runs counter to the traditional view that cancer progression is a linear process with new mutations building upon pre-existing ones.

The ultimate goal of cancer genomics is to give physicians the ability to tailor a patient's treatment regime not only to the type of cancer, but to the complete set of mutations exhibited by their specific tumour. Achieving this goal will require advances on many fronts, not least of which is a much more refined understanding of the relative contributions of 'driver' and 'passenger' mutations in the progression of cancer. At present, sequencing technologies are still not good enough to accurately and rapidly detect them all—and even if we could, in most cases we don't know how to use the data to make informed decisions about treatment. A related issue is the fact that tumours are very often heterogeneous; the genomes we sequence from them are an amalgamation of all of the genomes present in the sample, mutations included. Here the use of single-cell genomic and transcriptomic technologies presents new opportunities (see Chapter 6). In the case of hereditary cancers, the prospects are somewhat brighter, at least in so far as personal genome sequencing can help with screening and diagnosis.

Chapter 5
Evolutionary genomics

The *Oxford English Dictionary* defines 'palimpsest' as 'a manuscript or piece of writing material on which later writing has been superimposed on effaced earlier writing.' A genome is much like a palimpsest—underneath the current text can be seen traces of previous contributions from different 'authors' at different times. Although we can reasonably claim fluency in the language of the genome, we are still learning how to make sense of what genome sequences have to tell us. The genomic palimpsest provides a window on the evolution of life that cannot be opened using any other research avenue, and the field of comparative genomics has matured to the point that deep insights into some of the most fundamental questions in biology are within our grasp. Let's explore some examples, starting from the here and now and working further back in evolutionary time.

Molecules as clocks

The idea that molecules can be used as trackers of evolutionary history has been around a long time, longer in fact than techniques for sequencing DNA. In 1958, Francis Crick (of DNA-double-helix fame) envisioned a subject he called protein taxonomy:

> the study of amino acid sequences of the proteins of an organism and the comparison of them between species. It can be argued that these sequences are the most delicate expression possible of the phenotype of an organism and that vast amounts of evolutionary information may be hidden away within them.

These prophetic words capture the essence of the field known today as molecular evolution.

In early studies of haemoglobin, the oxygen transport protein found in vertebrate red blood cells, Linus Pauling and Émile Zuckerkandl noticed that sequence divergence increased with evolutionary divergence in a linear fashion. That is, the more distant the relationship between two organisms (as inferred from the fossil record) the greater the number of differences observed when their amino acid chains were aligned to one another. The same pattern was seen in other proteins such as mitochondrial cytochrome c (see Figure 11), raising the possibility that such data could be used to predict divergence times between organisms whose proteins had been sequenced but whose evolutionary relationship was unclear. With time it became apparent that different proteins had different evolutionary rates, but for a given protein the rate of change is roughly constant over time. This is the so-called 'molecular clock', a bedrock concept underlying modern comparative genomic research. Molecular clocks can be inferred using both protein (amino acid) and DNA (nucleotide) sequences, with the latter 'ticking' more quickly than the former. The rate at which non-coding DNA sequences such as introns and intergenic regions change over time is the fastest of all and is referred to as the neutral mutation rate.

Molecular clocks are far from perfect. Research has shown that DNA mutation rates can be influenced by many factors. External

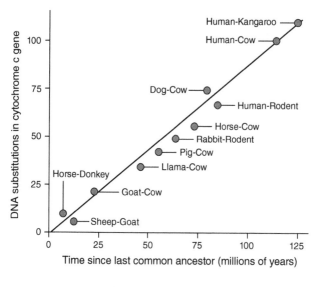

Human...	asn	leu	his	gly	leu	phe	gly	arg	lys	thr	gly	gln	ala	pro	gly	tyr	ser	tyr ...
Chimp...	asn	leu	his	gly	leu	phe	gly	arg	lys	thr	gly	gln	ala	pro	gly	tyr	ser	tyr ...
Whale...	asn	leu	his	gly	leu	phe	gly	arg	lys	thr	gly	gln	ala	val	gly	phe	ser	tyr ...
Fly...	asn	leu	his	gly	leu	phe	gly	arg	lys	thr	gly	gln	ala	ala	gly	phe	ala	tyr ...
Yeast...	asn	leu	his	gly	ile	phe	gly	arg	his	ser	gly	gln	ala	gln	gly	tyr	ser	tyr ...
Wheat...	asn	leu	his	gly	leu	phe	gly	arg	gln	ser	gly	thr	thr	ala	gly	tyr	ser	tyr ...

11. The molecular clock. Top: DNA sequence divergence levels between pairs of cytochrome c genes in animals plotted as a function of time (in millions of years) since divergence from a last common ancestor. Bottom: A portion of cytochrome c protein sequences from different organisms. Each block corresponds to a single amino acid (represented by a three-letter code, e.g. asn = asparagine, gly = glycine). Grey boxes indicate amino acids that differ from the human sequence.

forces such as radiation and chemical mutagens are known to cause DNA damage, but mutation is in fact a normal, inevitable part of genome biology. Intrinsic mutation rates can vary significantly from lineage to lineage depending on, for example, the fidelity of DNA replication and the efficiency of the cell's DNA repair mechanisms. Mutation rates can even differ between genomes residing in different compartments of the same cell, such as in the mitochondrion and nucleus of eukaryotes. This is because genome integrity is maintained by different subcellular machinery in different organelles, and they are exposed to slightly different biochemical environments (e.g. mitochondrial DNA is exposed to harmful reactive oxygen species generated on-site by aerobic respiration).

When looking at coding regions of a genome (i.e. protein-coding genes and genes for structural RNAs such as ribosomal RNAs), sequence divergence rates are heavily constrained by what is tolerated in terms of structure and function; natural selection will weed out organisms whose sequences have changed to the point that reproductive success is compromised. And of course molecular clock-based evolutionary inferences are only as good as the fossil calibrations to which they are linked. As new data emerge, researchers are constantly retesting and refining their methods to more accurately capture the complexities of the molecular evolutionary process. A well-calibrated molecular clock used for the purpose for which it was intended can serve as an invaluable framework for estimating when in deep time particular lineages of organisms emerged and the evolution of biological traits.

Population genomics of Adam and Eve

One of the first and most contentious applications of a DNA-based molecular clock was to better understand the evolution of our own species. Here the bacterial-like DNA found in the mitochondrion has proven extremely useful. Recall that the Y chromosome is unique to men, and mitochondria, though present in both males and females, are maternally inherited. Consequently, the probability that the mitochondrial and (if present) Y-chromosomal genotypes of any specific individual will persist for extended periods of time in a population is heavily influenced by the random nature of genealogy. Researchers have coined the terms 'Y-chromosome

Adam' and 'Mitochondrial Eve' to refer to the most recent common ancestors of these two distinct genetic components of the human genome, and, as we shall see, comparative genomics has been used to trace these ancestors back in time. Counterintuitive though it may seem, Adam and Eve need not have existed at the same time or in the same place.

The age of Mitochondrial Eve has remained relatively stable since it was first predicted in a landmark study published in 1987 by the New Zealand-born University of California (Berkeley) geneticist Allan Wilson and colleagues. Wilson's team used a molecular-clock approach to estimate that the mitochondrial DNAs of all modern humans trace back to a single woman living in Africa roughly 200,000 years ago. Using similar methodologies, Y-chromosome Adam was initially predicted to be a good deal younger: the common ancestor of the non-recombining portion of the Y-chromosome passed from fathers to sons was inferred to have lived ~100,000 years before the present. With the accumulation of more genomic data from geographically diverse males, however, Adam's age has been pushed further back to between 200,000 to 300,000 years ago.

These numbers do not mean that at specific points in time humanity was represented by a single living man and/or woman. Y-chromosome Adam and Mitochondrial Eve were simply individuals within their respective human populations to which all of the Y chromosome sequences and mitochondrial DNAs of present-day humans happen to trace. Maternally inherited mitochondrial DNAs and, in males, paternally derived Y chromosomes must obviously be passed on with each generation, but the genotype of any single human being is subject to extinction, the chances of which are elevated as a result of population bottlenecks (which are known to have occurred at various times during human evolution) (see Figure 12). The further back in time one goes, the number of ancestral genotypes that modern human genotypes can conceivably trace back to

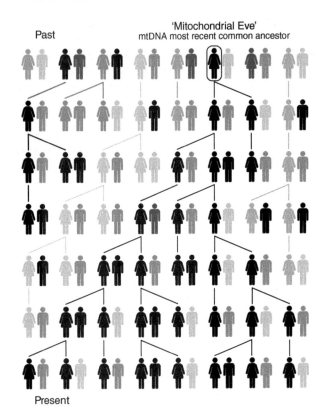

Past

'Mitochondrial Eve'
mtDNA most recent common ancestor

Present

12. **Maternal inheritance of human mitochondria and mitochondrial DNA (mtDNA). Black shows the maternal line that traces from the present back seven generations to the mtDNA most recent common ancestor. This single female, estimated to have lived in Africa ~200,000 years ago, is referred to as 'Mitochondrial Eve'. Various shades of grey correspond to maternal lines that have gone extinct due to random genetic drift or natural selection. Males (different shades of grey) inherit their mtDNA from their mothers but do not pass it on to their offspring.**

becomes smaller and smaller. At any point in time, there can be only one Y-chromosome Adam and Mitochondrial Eve, and their genealogies appear to be very different. The ages and identities of Adam and Eve are in fact not fixed but will change as human populations continue to evolve, and different maternal and paternal lineages go extinct.

Our understanding of Y-chromosome and mitochondrial-DNA evolution continues to be refined in response to new information. Of particular note is the fact that in humans and other animals, there appear to be exceptions to the rule of strict maternal inheritance of mitochondria. Even small and sporadic paternal contributions to present-day mitochondrial genotypes influence how researchers model the genealogical process and thus have the potential to confound molecular clock-based estimates of the age of Eve. And as vast quantities of human population genomic data continue to accumulate, researchers have begun the daunting task of comparing and contrasting the genealogical histories of mitochondria and Y chromosomes with those of the rest of the human genome—the sexually recombining autosomes, which we inherit from our mothers *and* fathers. What is emerging is a complex genomic mosaic across space and time, but one that is nevertheless consistent with the 'Out of Africa' model of human evolution. The picture has been complicated by evidence indicating that early humans interbred with Neanderthals and other extinct lineages after they evolved in Africa over 100,000 years ago and began to spread out across the globe.

Ancient DNA—from bones to genomes

Geologically speaking, the Holocene epoch is our epoch—it stretches from the present back ~11,700 years to the end of the last ice age. The study of organisms living prior to the Holocene falls under the domain of palaeontology, a field that relies heavily on interpretation of the fossil record. Advances in molecular biology and genomics have made it possible to study the genetic material of extinct organisms, providing exciting new avenues for reconstructing evolutionary history in concert with morphology-based investigations.

The field of ancient DNA research began taking shape in the 1980s, concomitant with—and indeed made possible by—the development of a laboratory technique called the Polymerase Chain Reaction (PCR). Invented by the American Nobel laureate Kary Mullis, PCR uses a polymerase enzyme, a pair of short DNA 'primers' (needed by the polymerase to initiate DNA synthesis), and dNTPs to exponentially amplify a DNA fragment of interest from a complex DNA sample. PCR is used in a wide range of applications, everything from paternity testing and crime scene investigations to the study of microbial ecology (see Chapter 6). For ancient DNA research, what makes PCR so powerful is that it allows one to study the tiny amounts of DNA that linger in biological remains such as desiccated tissue, teeth, and bone. Once amplified using PCR to manageable quantities, ancient DNA can be sequenced and compared to that of present-day organisms using standard bioinformatic techniques. Ancient DNA sequences are the molecular equivalent of a time machine; they can teach us about the biology of organisms that no longer exist.

The biggest challenge facing scientists studying ancient DNA is sample degradation—the older the fossil, the more fragmented its DNA is likely to be. Living cells contain nucleases, which are RNA- and DNA-cleaving enzymes whose activities are normally kept tightly in check. Upon death, however, these nucleases start to work indiscriminately, and because the cell's DNA repair mechanisms are no longer operational, the DNA is broken into smaller and smaller pieces. Such DNA also becomes chemically modified over time, which can inhibit the DNA polymerases used for sequencing as well as lead to the incorporation of incorrect nucleotides (and thus sequencing errors). The rate at which DNA degrades depends greatly on the environment in which an organism has come to rest. Generally speaking, DNA will persist longer in tissues that have been frozen or dehydrated quickly upon death, conditions that inhibit the activity of nucleases from the cells of the deceased individual as well as those of microorganisms living nearby.

Like many areas of cutting-edge science, ancient DNA research has had its share of controversy. Extraordinary measures must be taken to ensure that ancient DNA sequences are truly derived from the organism of interest, rather than from the person who extracted the DNA, from the palaeontologists who handled the fossil, or from the zoo of microbes on and within the fossil at the time it was collected (these microbes can themselves be ancient or modern). The sensitivity of PCR is such that even trace amounts of modern DNA contaminating lab equipment and reagents can yield PCR products that can be mistaken for ancient ones. Even airborne DNA can be amplified and sequenced if it finds its way into PCR reaction vessels. Early published accounts of ancient DNA sequences derived from insects trapped in amber, fossilized plants, and even dinosaur bones are widely believed to represent contamination from modern-day organisms. Ancient DNA research is now typically carried out in 'clean rooms' that have been sterilized with bleach and UV irradiation in order to minimize the chances of contamination. The upper time limit for the persistence of ancient DNA appears to be around one million years, and this is only in cases where the organism has been frozen.

Growing pains aside, *bona fide* ancient DNA sequences have been retrieved from diverse organisms across an impressively wide swath of time. For example, mitochondrial-DNA sequences have been retrieved from 1,000-year-old Egyptian mummies and from 'Ötzi', the ~5,200-year-old Iceman discovered in 1991 frozen in a glacier high in the Tyrolean Alps. Analyses of such sequences have provided insights into patterns of human migration and the origins of mitochondrial genotypes in present-day humans. As we will discuss in Chapter 7, gene fragments and even complete nuclear genomes have been sequenced from the frozen remains of woolly mammoths dating back 40,000+ years as well as ~300,000-year-old cave bears. Further back still, plant and animal sequences have been PCR-amplified from ice cores in Greenland dating to ~800,000 years before present. The oldest nuclear genome sequenced thus far comes from a ~700,000-year-old wild horse bone frozen in north-western Canada's Yukon Territory.

The Neanderthal genome

Among the most remarkable achievements of ancient DNA research has been the sequencing of nuclear genomes from Neanderthals and other now-extinct human relatives. The first such genome was published in *Science* in 2010, work spearheaded by one of the field's principal architects, Svante Pääbo, a Swedish scientist at the Max Planck Institute for Evolutionary Anthropology in Germany. Throughout the 1980s and 1990s, Pääbo carried out pioneering studies on the persistence of DNA in human mummies, extinct sloths, and cave bears, and in 1997 led the first successful effort to sequence fragments of mitochondrial DNA from a Neanderthal bone dating back ~40,000 years. The data provided empirical support for the now generally accepted hypothesis that all modern humans originate from Africa.

Pääbo was one of the first to recognize that next-generation sequencing had the potential to transform ancient DNA research in the same way that PCR had. As discussed in Chapter 2, a limitation of early next-generation technologies was short sequence read lengths, initially on the order of fifty nucleotides. Depending on the size and complexity of the target genome, this can pose a serious challenge for accurate genome assembly. However, given that ancient DNA extracts are already highly fragmented, read length becomes much less of an issue. What matters most is throughput, and next-generation sequencing platforms produce so much data that it is possible to obtain satisfactory levels of coverage even for large animal genomes.

And in cases where a high-quality reference genome has already been sequenced from a close relative, *de novo* genome assembly is unnecessary (see Chapter 3); ancient DNA sequence reads are simply 'mapped' onto the reference, and the nucleotide differences between the ancient and modern genomes can be determined. This is the strategy that was used to sequence the Neanderthal genome, with the human genome as a scaffold, and in the case of the extinct woolly mammoth, where reads were mapped against the nuclear genome sequence of an Asian elephant (see Chapter 7).

The Neanderthal genome project was a great technological and bioinformatic feat. Pääbo's team first published one million bp of Neanderthal DNA sequence in 2006, obtained using pyrosequencing in collaboration with 454 Life Sciences. This benchmark study showed that it was in principle possible to completely sequence a large nuclear genome from an ancient DNA sample, but also that a brute force approach would be needed. Only rarely does more than 1 per cent of the DNA extracted from a fossil actually come from the fossil; the researchers had to investigate many different Neanderthal bones in order to find ones that would yield sufficient starting material for sequencing (not to mention convince museums to part with their precious fossils). They settled on bones from three different individuals, ~40,000–50,000 years old, found in Croatia's Vindija Cave. Procedures for the production of DNA libraries for sequencing also needed to be modified so as to minimize the loss of DNA. The genome was eventually completed using Illumina, a 'sequencing-by-synthesis' technology that produces shorter individual reads than pyrosequencing but a much higher total number of reads per instrument run. From many hundreds of runs, enough Neanderthal-derived reads were obtained such that the genome could be covered to a depth of ~1.3x (a total of >4 billion base-pairs). More than 95 per cent of the raw sequence reads were ultimately discarded as contamination. To help distinguish *bona fide* ancient reads from modern human ones,

Pääbo and colleagues sequenced the genomes of five individuals living in different parts of the world. All Neanderthal and human reads were then mapped against the chimpanzee genome; being much more distantly related to both humans and Neanderthals, the chimpanzee comparison served as a critical reference point for quantifying the Neanderthal-human sequence divergence, as well as a control for contamination (chimps do not collect fossils and are not employed in sequencing facilities!).

Overall, the Neanderthal and human genomes were found to be ~99.5 per cent identical, compared to ~99 per cent for the chimp-human comparison. The most stunning discovery was that genome sequences from three non-African populations (Chinese, Papua New Guinean, and French) were ~4 per cent more similar to the Neanderthal genome than were the genomes of Western and Southern Africans. This small but clear Neanderthal 'footprint' was interpreted as evidence for interbreeding between Neanderthals and the early ancestors of present-day non-African populations (see Figure 13), perhaps in the Middle East where they appear to have coexisted for a time. The evidence supporting this controversial hypothesis, which has been around for several decades, has gotten stronger with the sequencing of several additional ancient genomes, and their increased quality has made it possible to extract information that could not be gleaned from the original low-coverage sequence. For example, a 2014 *Nature* publication described a Neanderthal genome sequenced to a depth of 50x, on par with that routinely achieved for human genomes. This particular genome, derived from a ~50,000-year-old fossil found in Denisova Cave in the Altai Mountains of Siberia, was found to contain long tracts of homozygosity (stretches of sequence in which maternal- and paternal-derived chromosomes are the same) and a genome-wide level of heterozygosity much lower than that seen in modern-day humans. This indicates that this individual, a female, belonged to a small population and had parents who were very closely related (i.e. half-siblings or equivalent).

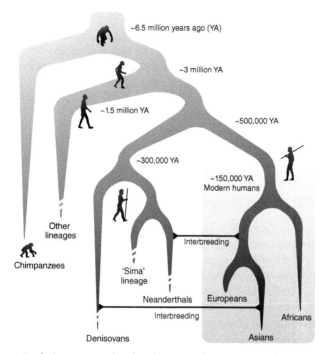

13. Evolutionary tree of modern humans and their closest relatives. Diagram shows two documented instances of interbreeding between the ancestors of modern human lineages and extinct ancestors, one involving Neanderthals and another the so-called Denisovans. The 'Sima' lineage refers to a Neanderthal-like genome sequenced from a sample taken from Spain's Sima de los Huesos.

One additional lineage for which ancient genome sequence data have been obtained is worthy of mention: the Denisovans. In collaboration with Russian scientists, Pääbo's team extracted DNA and sequenced a nuclear genome to 30x coverage from a tiny, exceptionally well-preserved (but undated) finger bone found in the Denisova Cave, mentioned earlier. Not surprisingly, the bone itself reveals little about its owner. The genome sequence, however, indicates that it is not of classical

Neanderthal ancestry, but rather a member of a distinct branch of early human evolution that diverged from Neanderthals after modern humans did (see Figure 13). Like Neanderthals, the Denisovans appear to have contributed DNA to modern humans, albeit to different populations (in this case, specifically with early Melanesians) and after Neanderthals did. And like the Altai Neanderthal genome, the specific features of the Denisovan sequence indicate that the mystery bone comes from a member of a small, declining population with a low level of genetic diversity.

What makes us human?

Precisely when, where, how, and how much the ancestors of modern humans interacted with Neanderthals, Denisovans, and other archaic lineages are important questions for the future. Given the reality of DNA degradation, it remains to be seen how much further back in time we can go in our quest to understand the evolution of our species. Scientists are of course eagerly trying to find out. For example, nuclear genomes have recently been sequenced from early modern human bones ~40,000 years of age found in Siberia and Romania. The 'Romanian' genome provides the strongest evidence yet for Neanderthal-human interbreeding; this individual appears to have had a Neanderthal ancestor only four to six generations back. Researchers have also managed to coax a Neanderthal-like genome sequence from severely damaged DNA extracted from remains in Northern Spain's famous Sima de los Huesos ('pit of bones') (see Figure 13). These fossils are ~430,000 years old. The tools of the ancient DNA trade are now in place to make use of fossils going back a million years. And with the data already in hand, experts have begun to address the question of what distinguishes modern humans from our closest extinct relatives. Of the hundreds of thousands of DNA substitutions that occurred along the branch leading to us, which ones changed our biology in ways that impacted our evolutionary 'success'?

Comparisons of Neanderthal and Denisovan genomes to those of chimps and humans show that while ~3,000 human-specific substitutions occur in regions of the genome predicted to regulate gene expression, only ~100 substitutions actually result in an amino acid change (i.e. a phenotypic change). At present, the functions of only a handful of these modified proteins are sufficiently well understood so as to be informative, but among them is a protein involved in sperm motility and two others associated with skin. Recent analyses of the wealth of data emerging from the 1000 Genomes Project have shown a non-random distribution of Neanderthal-derived DNA across human chromosomes. In general, gene-rich regions of the human genome tend to have fewer Neanderthal-type alleles, and a striking five-fold reduction in Neanderthal DNA was observed on the X chromosome. (The term 'allele' is used in reference to different versions of the same stretch of DNA.) Furthermore, genes that are very highly expressed in human testes (and not in other tissues) are also largely devoid of Neanderthal-derived alleles. These results are consistent with the idea that although Neanderthals and the ancestors of modern humans clearly interbred, at least some of their offspring were reproductively disadvantaged. Conversely, genes enriched in Neanderthal-type alleles include those encoding important proteins in hair and skin, notably keratin, suggesting that acquisition of these stretches of DNA played a role in human adaptation to colder environments outside Africa.

Evolution in action

A large body of evidence indicates that natural selection has acted upon genes associated with our diet, pigmentation, resistance to pathogens, and various aspects of our physiology that impact our fitness in different environments. A well-studied example is the *LCT* locus, which encodes the lactose-digesting enzyme lactase. As mammals, the ancestral state for humans is 'lactase non-persistence', in other words loss of the ability to digest lactose after weaning. However, in European and certain East

Genomics

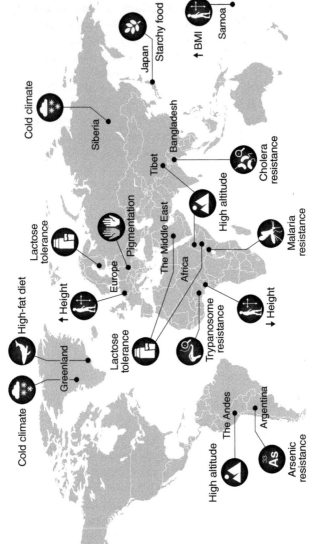

14. Global distribution of adaptive traits in humans inferred from genomic data (map of globe designed by Freepik).

African populations, the region of the genome surrounding *LCT* shows evidence of past selective pressure for the ability to use milk as an energy source into adulthood, and laboratory experiments confirm that the genetic variants linked to lactase persistence contribute to increased *LCT* expression. Large tracts of *LCT*-associated homozygosity present in modern Europeans and East Africans, but largely absent in ancient genome sequences, are indicative of selective sweeps occurring ~9,000 years ago for the European *LCT* variant and ~5,000 years ago for the East African one (see Figure 14). These estimates align neatly with archaeological data pointing to the domestication of cattle in the Middle East and Northern Africa beginning ~10,000 years before present.

Another example of human evolution in action is variation in the number of genes encoding salivary amylase (*AMY1*), the enzyme that digests starch into simple sugars. Research has shown a direct relationship between *AMY1* copy number and the amount of amylase protein in our mouths, and while chimps have a single *AMY1* gene, we humans can have a half-dozen or more. Interestingly, individuals from populations with 'high-starch', agriculture-based diets, such as the Japanese, tend to have more *AMY1* genes in their genomes than do those with 'low-starch' diets such as traditional hunter-gatherers. This has been touted as an example of positive selection, whereby gene duplications led to increased *AMY1* gene copy number and increased amylase protein, which conferred a fitness benefit by allowing more efficient extraction of energy from starch.

Additional genomic regions associated with recent diet-related episodes of selection include the *CREBRF* and *FADS* loci. An amino acid-changing mutation in *CREBRF*, the gene for an important cell-signalling protein, is associated with increased body mass index (BMI) in the Samoans of Polynesia (see Figure 14). Although today linked to an increased risk of obesity, genomic data suggest that in the recent past this high-frequency allele was positively selected for its ability to confer resistance to starvation in the Samoan founder population. In the case of *FADS*, a region of chromosome 11 that encodes fatty acid desaturase enzymes, comparisons of the Inuits of Greenland, Native Americans, and other indigenous and non-indigenous peoples show striking variation in allele frequencies consistent with an adaptive event linked with metabolic adaptations to a high-fat diet in the Arctic. Evidence suggests that this took place in Beringia—site of the prehistoric land bridge across what is now the Bering Strait—where humans lived for many thousands of years after first migrating east from Asia around the time of the last ice age. FADS variants eventually spread south with the indigenous peoples who first populated the Americas.

The diversity of skin pigmentation in present-day humans can be seen as a trade-off between protection against the damaging effects of ultraviolet (UV) radiation and the maintenance of vitamin D synthesis, which is a process that requires UV exposure. Research indicates that selection pressures have influenced the colour of our skin in both directions. A recent scan of 230 'ancient' Eurasian genomes (i.e. from individuals who lived between 300 and 6,500 BC) showed that genetic variants associated with light skin pigmentation shifted from low frequency ~4,000 years ago to very high frequency in modern-day Europeans (interestingly, this scan also revealed evidence for directional selection for height in certain European and African populations; see Figure 14). One important gene is *MCR1*, which encodes a protein that helps stimulate skin cells to produce melanin. While Africans have only a single *MCR1* allele, dozens of variants are found in East Asian and Caucasian populations; at least five of these variants give rise to amino acid changes that result in the loss of protein function, and many appear to be of Neanderthal ancestry. Selection thus appears to have acted strongly against *MCR1* mutations in humans living in high-UV environments near the equator, with skin-lightening mutations being tolerated in more northerly regions. Another gene known to influence skin colour is

SLC24A5, mutations in which give rise to proteins with different amino acid sequences that also contribute to pale skin. Additional pigment-associated loci are emerging from genetic studies of skin and eye colour in the island country of Cape Verde, whose present population has been influenced by genetic mixing between ancestral African and European populations over the past few hundred years.

Genome-wide scans for evidence of natural selection have also provided insight into how humans have adapted to life in extreme environments. For example, genomic comparisons of people living in three high-altitude locations—the Tibetan Plateau, the Andes Mountains of South America, and Ethiopia's Semien Plateau—with their closest low-altitude relatives have revealed signatures of positive selection at genes whose expression is induced under low-oxygen conditions (see Figure 14) (intriguingly, the adaptive haplotype found in Tibetans appears to be of Denisovan ancestry). With respect to temperature, a transcription factor gene (*TBX15*) has been shown to play a role in the differentiation of certain types of fat cells that, upon exposure to cold temperatures, begin to produce a protein (thermogenin) that helps generate heat. The *TBX15* region of the genome shows signatures of positive selection in the Inuit, and has thus been linked to cold adaptation.

Genomic adaptation to the toxic element arsenic has been found in members of the town of San Antonio de los Cobres in north-western Argentina. Arsenic exists here naturally at levels ~10× that considered safe by the World Health Organisation. Humans have nevertheless lived in the region for >10,000 years, and its present-day inhabitants possess variant forms of genes known to function in arsenic metabolism, variants that are present at a much lower frequency in their closest relatives living in areas with low arsenic levels.

Finally, genomic investigations have shown evidence for selection in populations with regular exposure to pathogens such as the malaria-causing protist *Plasmodium* (transmitted by mosquito bites), the trypanosomes that cause African sleeping sickness, and the bacteria that cause cholera. From an evolutionary perspective, such examples are interesting in that the specific alleles associated with pathogen resistance can themselves also be linked to disease. For example, the genetic variants that confer resistance to malaria can also give rise to sickle cell anaemia and other blood disorders. In pathogen-exposed populations, these disease-associated alleles are thus actively maintained in the gene pool.

Animals and multicellularity

Thus far we have focused on examples in which evolutionary inferences are based on the analysis of DNA. Although powerful, this approach has its limits. The molecular clock 'ticks' rapidly for nucleotide sequences, especially in non-coding DNA, and combined with the effects of genome rearrangement, it can be difficult to identify homologous genomic regions between distantly related organisms. The genes themselves can also become 'saturated': if enough time has passed, substitutions will have occurred multiple times in the same nucleotide position after species divergence, which serves to erase evolutionary signal and add 'noise'. When saturation is known or suspected, researchers turn to the analysis of protein (i.e. amino acid) sequences, which, due to the universality of the genetic code, can be accurately inferred from gene sequences. Combined with the protein-based construction of phylogenetic trees, the gene repertoires of diverse organisms can be compared in order to reconstruct important events in evolution. The origin of multicellularity is a fascinating case in point.

For one and a half to two billion years after life first arose, organisms were exclusively microbial. True multicellularity is the domain of eukaryotes; beyond the obvious examples of animals and plants, multicellular organisms have evolved from single-celled ancestors more than a dozen times independently over the past 800 million years, within the fungi and in seaweeds such as red algae and kelp, for example. Genomics has enabled such insights in two ways. First, an analytical approach called 'phylogenomics' has made it possible to infer relationships between anciently diverged organisms by combining the sequences of many different proteins together in a single analysis. This is necessary because the amount of information that can be extracted from a single protein is often limited, and analysing the relationships among multiple sequences at once increases the signal-to-noise ratio. Phylogenomic analyses of hundreds of proteins simultaneously are now routine and have provided researchers with a 'road map' of eukaryotic relationships. Second, mapping the presence and absence of key genes associated with multicellularity onto the tree of eukaryotes has provided insight into when and how each gene arose in each of the different groups. While some common themes have emerged, the take-home message is that the molecular systems underlying multicellularity differ from group to group. Evolution works with what is available.

In the case of animals, some components of the multicellularity 'toolkit' are truly ancient. Phylogenomic analyses have shown that animals share most recent common ancestry with fungi (not plants as once believed), and branching sister to well-known animal groups such as chordates, arthropods, and sponges are several unicellular lineages whose members are capable of forming colonies and cellular aggregates (see Figure 15). Best studied among them are the choanoflagellates ('collared flagellates'), whose cellular structure is reminiscent of the feeding cells of sponges. Despite not being animals, the genomes of choanoflagellates, as well as those of two other single-celled 'holozoans' (i.e. animals plus their closest relatives), have genes that encode signature animal proteins involved in cell–cell adhesion and signalling (e.g. cadherins, lectins, and focal adhesion tyrosine kinases). These animal-associated molecular features thus arose during a phase of evolution long before the origin of complex animal multicellularity and the subsequent Cambrian

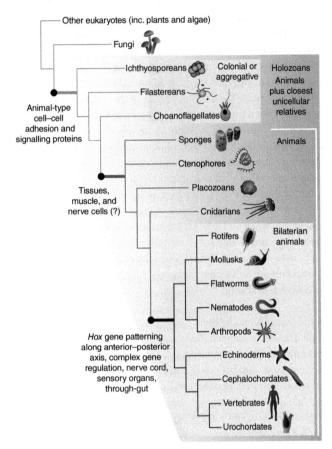

15. Evolution of animals and their closest unicellular relatives.

explosion; gene duplication and functional differentiation of these and other ancestral genes played a key role in the rise of complex animal life.

A more recently evolved example is the homeotic genes, or *Hox* genes. Originally discovered in fruit flies, *Hox* genes encode

DNA-binding transcription factors that control development along the head-to-tail axis of bilaterally symmetric animals. Comparative genomics has shown that while most invertebrates possess a single cluster of *Hox* genes, mammals and fishes can have four or more clusters. Remarkably, experiments in various animals systems show that the temporal-spatial expression of *Hox* genes during development typically mirrors their position on the chromosome: genes at the end of the *Hox* cluster are turned on first in the anterior parts of the embryo, while genes at the beginning of the cluster are expressed later on in posterior regions. This *Hox* gene-based developmental programme was clearly hard-wired into the biology of the earliest bilaterally symmetric animals (bilaterians) (see Figure 15).

The evolution of other hallmark animal features such as muscles and the nervous system is less clear. In molecular analyses, the ctenophores (comb jellies), which possess rudimentary muscles and nerve cells, are sometimes resolved as the most deeply diverged of all true animals, suggesting that these features evolved well before the first bilaterians arose 500–600 million years ago. However, in a recent phylogenomic analysis of 1,719 proteins designed to minimize tree reconstruction artefacts, the sponges were found to branch deepest. Although sponges are considered to have tissues, they lack true muscle and nerve cells; these features may therefore have evolved later in a common ancestor shared by ctenophores, cnidarians (e.g. jellyfish), placozoans, and bilaterians (see Figure 15). The placozoans ('flat animals') are a particularly troublesome lineage in this regard. Although superficially amoeba-like in appearance, placozoans are clearly animals in terms of their multicellularity and phylogenomic position—they may have lost muscles and nerves secondarily. This is a conundrum that often arises when molecular and morphological data are compared, underscoring the importance of having an evolutionary framework upon which to map organismal characteristics.

Eukaryotes and organelles

Next to the origin of life itself, the question of how eukaryotic cells evolved from prokaryotes is one of the most difficult in all of biology. Genome sequences speak strongly to the distinctness of Bacteria (or Eubacteria) and Carl Woese's Archaea (Archaebacteria), and show that the Eukaryota are in essence a hybrid of the two (see Figure 16). Eukaryotic genes associated with cellular metabolism and membrane biology are generally of bacterial ancestry, while 'housekeeping' genes involved in information processing (e.g. protein synthesis, transcription, and DNA replication) are derived from archaea. Of the eukaryotic proteins that have clear sequence similarity

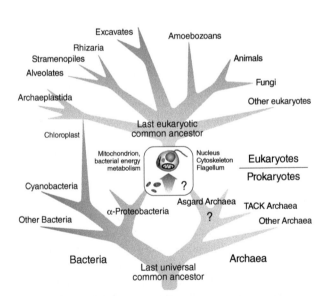

16. The Tree of Life. Diagram highlights the evolutionary origin of the eukaryotic cell from the α-proteobacterial progenitor of the mitochondrion and an archaeal host, perhaps related to an ancestor of the so-called Asgard Archaea.

to those of prokaryotes (and there are many that do not), 'bacterial' proteins outnumber archaeal ones approximately three-fold. The reason(s) for this imbalance are unclear. Even after several decades of study, the prokaryote-to-eukaryote transition remains mysterious.

The first archaeal genomes were sequenced in the 1990s, analyses of which supported the so-called 'three-domains' view of life: archaea and eukaryotes are evolutionarily distinct sister domains to the exclusion of bacteria. However, when universally distributed proteins were subjected to phylogenomics, eukaryotes sometimes emerged from *within* the archaea. This idea was first proposed in 1984 by UCLA evolutionary biologist James Lake who, remarkably, noted morphological similarities between the ribosomes of eukaryotes and certain archaea under the electron microscope.

The notion that eukaryotes might have arisen from within an already diversified archaeal lineage has been bolstered with the recent discovery of novel archaeal groups in sediments taken from extreme environments such as 'Loki's Castle', a deep-sea hydrothermal vent field located between Greenland and Norway. Found by a team of scientists led by Thijs Ettema at the University of Uppsala, these so-called Asgard archaea are named after Norse gods—Lokiarchaeota, Thorarchaeota, and so on. Their genomes are a curious mix of genes for 'standard' archaeal proteins and 'eukaryotic signature proteins' (ESPs), proteins originally assumed to be unique to the inner workings of eukaryotes. These ESPs include proteins that in eukaryotic cells form the rigid cytoskeleton and transport vesicles from A to B. The Asgard archaea thus have the potential to transform our understanding of the elusive host cell that partnered with the bacterial progenitor of the mitochondrion. Unfortunately, everything we know about their biology has been inferred from genomes that have been stitched together from sequences obtained directly from environmental DNA by metagenomics (see Chapter 6); the actual cells have never been seen, let alone cultured or studied in the lab. The functions of the ESPs in Asgard taxa are thus unknown.

Hypotheses for the evolution of eukaryotes are of two general sorts. The 'mitochondria-late' model has it that the salient features of eukaryotic cell biology, such as the nucleus, endomembrane system, and cytoskeleton, evolved prior to the endosymbiotic origin of the mitochondrion. Here the ability to ingest prey (using the cytoskeleton) is seen as an essential prerequisite for mitochondrial evolution. In contrast, the 'mitochondria-early' scenario posits that the mitochondrion evolved in the context of a prokaryotic host and that the eukaryotic cell could not have arisen without the evolution of this textbook organelle. With their ESPs, the Asgard taxa could represent 'missing links' between prokaryotes and eukaryotes, but at present there are not enough data to distinguish between the two models.

What we do know is that the mitochondrion evolved from a member of the α-Proteobacteria and that the last common ancestor of all known eukaryotes had such an organelle (see Figure 16). Although consistent with 'mitochondria-early', this does not preclude the possibility that mitochondrion-lacking eukaryotes once existed but went extinct, or that such organisms have simply escaped our notice. Indeed, an unassuming anaerobic protist named *Monocercomonoides* (which dwells happily in chinchilla intestines) shows that it is possible to live mitochondrion-free. Genes of mitochondrial ancestry were detected in its nuclear genome, but not a single trace of genes for mitochondrion-targeted proteins was found.

It is important to recognize that *Monocercomonoides* is *not* an 'ancient' eukaryote: it is of course alive today and phylogenomics shows that it is specifically related to mitochondrion-containing 'excavate' protists such as the pond alga *Euglena* and trypanosomes (see Figure 16). Its mitochondrion was clearly lost secondarily. *Monocercomonoides* appears to have achieved this feat in part by replacing its genes for iron-sulfur (Fe-S) cluster biosynthesis—an essential biochemical process in all eukaryotes that occurs in the mitochondrion—with other versions acquired from bacteria. In *Monocercomonoides*, experiments show that Fe-S biosynthesis takes place in the cytoplasm. (A handful of other anaerobic protists have been shown to retain a mitochondrion but lack a genome; these genome-free mitochondria are, as expected, the site of Fe-S biosynthesis and other biochemical processes.) All things considered, the jury is still out as to whether the eukaryotic cell evolved concomitantly with the mitochondrion or whether the organelle evolved after the main features of eukaryotes had already arisen.

Tree of life or web of life?

All eukaryotic organisms are genomic and cell biological mosaics. The picture becomes even more complex when one considers the chloroplasts (or plastids) of photosynthetic lineages. Like mitochondria, chloroplasts have genomes that are tiny compared to those of prokaryotes, but the 50–200 genes they retain point clearly to bacteria—in this case, photosynthetic cyanobacteria (see Figure 16). Remarkably, the chloroplasts of both red and green algae have been passed to other eukaryotes by what is called 'secondary' endosymbiosis. Here the chloroplast-bearing endosymbiont brings with it a nucleus and mitochondrion whose genes meld with the nuclear genome of the host organism (in some cases, the nucleus of the endosymbiont persists, resulting in an organism with four or even five different genomes). Secondary endosymbiosis is not simply an evolutionary oddity: it gave rise to ecologically significant phototrophs such as bloom-forming diatoms and 'red-tide' dinoflagellates. Even examples of 'tertiary' endosymbiosis have been described (think of them as the cellular equivalent of a Russian nesting doll). Taking into account the evolution of mitochondria and chloroplasts, genes have moved both vertically and horizontally in eukaryotes since they first emerged ~1.5 billion years ago.

Beyond endosymbiosis, however, there is debate as to how much horizontal gene transfer (HGT) occurs in eukaryotes. HGT, the exchange of genetic material across species bounds, is well-known as a driver of innovation in prokaryotes, and three mechanisms have been described. Conjugation involves the passage of DNA between prokaryotic cells that are physically connected via a thin tube called a pilus. Transformation is when a prokaryote takes up foreign DNA that is free-floating in the environment, such as that released from dead cells. And transduction is genetic exchange mediated by viruses, which can pick up pieces of DNA from one cell and pass them on to another. The relative significance of these three forms of genetic exchange varies between and within different bacterial and archaeal lineages, but each has been shown experimentally and bioinformatically to be capable of moving DNA from lineage to lineage. Beyond viral-mediated gene transfer, no such mechanisms have been described in eukaryotes, even though comparative genomics has shown evidence for prokaryote-to-eukaryote and eukaryote-to-eukaryote HGT in protists, plants, and some animals.

Among bacteria and archaea, HGT is so pervasive that many scientists now consider the prokaryotic tree of life to be more like a web. Take *E. coli*, for which >2,000 genomes have been sequenced. As of this writing, the *E. coli* 'pan-genome' contains ~90,000 genes, in other words there are ~90,000 different genes that have been found in the genome of one or more *E. coli* strains, and the pan-genome continues to grow by thirty to fifty genes with each new genome sequenced. This is remarkable given that the number of genes in any given *E. coli* strain rarely exceeds 4,000. Many of these strain-specific genes facilitate adaptation to new environments and pathogenicity (e.g. the unfortunate differences between the diarrhoea-causing *E. coli* O157:H7 and the normal *E. coli* in our gut are the result of HGT-derived genes in the former). In contrast, the *E. coli* core genome—the set of genes found in *all* genomes of the species—is only ~1,000, and is getting smaller over time. If one considers all sequenced bacterial genomes, the core is <150 genes! Tree of life or web of life (or both), prokaryotic genomes are exceptionally diverse and dynamic. And as we shall see in Chapter 6, we have only glimpsed the tip of the diversity iceberg. Here too modern genomic tools continue to open new paths of exploration.

Chapter 6
Genomics and the microbial world

If one attempts to culture microbes from an environment such as soil or seawater, ~100 times more cells can be seen under the microscope than will yield colonies on a petri dish. This observation, known as 'the great plate count anomaly' (see Figure 17), has intrigued and frustrated microbiologists for decades. This is not to say that most microbes in nature *cannot* be cultured in the lab, just that attempts to do so have thus far been unsuccessful. The most readily cultivated organisms are, according to microbial ecologist Philip Hugenholtz, 'the "weeds" of the microbial world and are estimated to constitute less than 1% of all microbial species'. Together, molecular biology and genomics have made it possible to explore the diversity and ecology of the remaining 99 per cent.

Environmental gene sequencing

Molecular tools for probing the hidden microbial majority independent of culture were pioneered by the University of Colorado biochemist Norman Pace in the 1980s. The procedures revolved around PCR, the *in vitro* DNA amplification technique that uses DNA polymerase, a pair of DNA 'primers', and dNTPs to exponentially amplify specific DNA sequences from minute quantities of starting material (see Chapter 5). Like his mentor Carl Woese, Pace targeted the small subunit (SSU) rRNA gene, whose ubiquity and high degree of sequence conservation makes

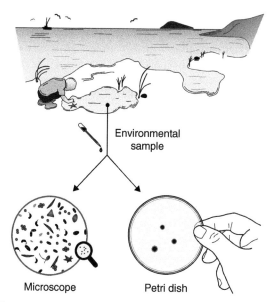

Environmental
sample

Microscope

Petri dish

17. The great plate count anomaly.

it particularly amenable to PCR amplification from 'community
DNA' (i.e. DNA extracted not from a single cultured organism
but directly from an environmental sample). Using PCR primers
designed to match the most conserved regions of the rRNA gene,
Pace and colleagues showed that gene fragments from a wide
range of microbes could be amplified in a single experiment. The
amplified DNA fragments are cloned, sequenced, and analysed
using various bioinformatic tools.

Over the past two decades, 'environmental PCR' has been used
to examine the composition of microbial communities in a wide
range of environments: seawater, marine sediments, solar
salterns, crop soil, tree bark, oil wells, even air conditioners,
shower curtains, and computer keyboards. As a result, our
knowledge of prokaryotic genetic diversity has increased
dramatically, both within and beyond previously recognized

groups of organisms. The precise numbers are still being debated, but the existence of a hundred or more bacterial phyla has been predicted (up from a dozen or so in the late 1980s), and a similar level of diversity seems possible within the archaea. The number of distinct prokaryotic rRNA genes that have been sequenced and deposited in public databases is now in the millions.

Environmental PCR has proven equally effective as a tool for exploring the diversity of microbial eukaryotes. Beginning in the early 2000s, culture-independent molecular surveys of marine habitats have greatly expanded knowledge of plankton diversity, leading to the discovery of several algal groups completely new to science. Of particular note is a comprehensive survey of marine micro-eukaryote diversity carried out as part of the recently completed Tara Oceans expedition. Over the course of several years, scientists aboard the French research schooner *Tara* collected thousands of water samples from around the globe and from a variety of depths. DNAs were extracted, environmental PCRs performed, and massively parallel next-generation sequencing techniques used to generate millions of DNA sequence reads. Beyond the astonishing levels of genetic diversity observed, researchers found that the relative abundances of key photosynthetic and non-photosynthetic lineages were very different from what had previously been assumed.

Closer to home, our understanding of the sea of microbes living in and on the human body has been transformed by the application of molecular methods. Environmental PCR surveys have shown the existence of hundreds of hitherto unrecognized bacterial species living in our mouths, in our belly buttons, and in our gastrointestinal tracts. Various anaerobic micro-eukaryotes and methane-producing archaea have also been identified; although much less abundant than the bacteria, these organisms appear to influence our gut physiology. The so-called human microbiome has become the subject of intensive study, with teams of physicians, microbiologists, and genomicists trying to understand how and why its composition changes over time, and how this impacts our health and well-being.

Metagenomics: engine of discovery

Environmental PCR is a powerful tool for exploring microbial diversity, but because it is (for the most part) focused on evolutionarily conserved marker genes such as rRNA, it provides researchers with very little direct information about the biology of the underlying organisms. It can help us answer the question of 'who is there?' but not 'what are they are doing?' Advances in molecular biology, genomics, and bioinformatics have moved the field well beyond the targeted amplification of specific genes from environmental samples.

The term 'metagenomics', first used in a 1998 publication by American microbiologist Jo Handelsman and colleagues, refers to the isolation and analysis of large DNA fragments taken directly from the environment. One of the key advantages of this approach, also known as environmental genomics, is that it avoids biases associated with PCR amplification; it allows scientists to obtain random samples of dozens to hundreds of genes residing on the same piece of genomic DNA, thereby providing insight into the biological properties of the donor organism. (Note that some researchers consider PCR-based, gene-specific surveys of environmental DNA to fall under the umbrella of metagenomics.)

A landmark demonstration of the promise of metagenomics was the 2000 discovery of a novel form of light-based energy generation used by microbes in the ocean. In analysing a large chunk of DNA isolated from a water sample taken in Monterey Bay, California, Oded Béjà, Edward DeLong, and collaborators found a gene encoding a rhodopsin protein. In and of itself, this discovery was not so unusual, as rhodopsins are found in a variety of different organisms; they function as light sensors in eukaryotes (including the eyes of humans and other animals), and as light sensors and ion/proton pumps in salt-loving archaea. In this case, however, bioinformatic investigation showed that the rhodopsin gene was located on genomic DNA that was very clearly of bacterial origin, more specifically, from a proteobacterium. The new protein was dubbed 'proteorhodopsin', and subsequent research has shown that a variety of marine bacteria, and even a handful of single-celled eukaryotes, use membrane-associated rhodopsins to capture solar energy. The spread of rhodopsin genes to such a wide range of microorganisms appears to have been facilitated by horizontal gene transfer (see Chapter 5).

The proteorhodopsin system is very different from the chlorophyll-based light-capturing mechanism used by cyanobacteria and algae, and its discovery changed the way scientists think about the flow of energy through microbial ecosystems. There is still much to learn. Massive proteorhodopsin gene diversity was found in metagenomic data derived from surface water samples all over the world as part of Craig Venter's 2003–4 Global Ocean Sampling Expedition. Beyond proteorhodopsins, this single metagenomic study identified more than six million distinct protein sequences, more than double the number present in public databases at the time. Almost 2,000 of these proteins showed no obvious sequence similarity to previously characterized proteins. We still do not know what most of these proteins do and, in many cases, which types of organisms they belong to.

Our understanding of the nitrogen cycle has been similarly transformed by metagenomics. The process of aerobic nitrification—the conversion of ammonia to nitrate via nitrite—was long thought to be unique to certain bacteria. But in the mid-2000s, researchers found genes for ammonia oxidation on large chunks of DNA belonging to archaea. Ammonia-oxidizing archaea have since been found in diverse terrestrial and aquatic habitats, including coastal waters (where they sometimes live symbiotically with sponges) and samples taken from the deep sea. In some environments, ammonia-oxidizing archaea are more abundant than their bacterial counterparts. Metagenomics is thus helping to fill in gaps in our knowledge of the biogeochemical cycles that sustain life on Earth, gaps that in many cases we didn't know existed. And with the burgeoning fields of 'metatranscriptomics' and 'metaproteomics', researchers are now able to move beyond genotype to phenotype, exploring the diversity and abundance of environmental RNA and protein, respectively.

Complete microbial genomes from the environment

Critical to the success of metagenomics is the ability to propagate large chunks of DNA in the lab. Tool selection is important. Most plasmids, such as those used in Sanger sequencing, are only useful for cloning small DNA fragments, typically 1–5 Kbp. However, specialized cloning vectors such as 'bacterial artificial chromosomes' allow the stable insertion and propagation (usually in *E. coli*) of pieces of DNA up to 300 Kbp. Even larger inserts are possible using so-called 'yeast artificial chromosomes' (YACs) in the model eukaryote *Saccharomyces*: with this system, chromosome-sized pieces of DNA up to ~2,000 Kbp in size can be cloned. (YACs were used during the initial 'mapping' phase of the publicly funded Human Genome Project, although their ultra-large inserts were sometimes unstable and difficult to analyse with confidence.) Assuming one can extract high-quality DNA from an environmental sample, large-insert vectors can be used to construct a metagenomic library, a repository of all of the genomic information that was present in that sample. Such libraries are no different than those used in traditional genome sequencing projects, and the DNA within them can be queried using the same tools.

For example, a metagenomic library can be screened in a systematic fashion for the presence of specific genes by PCR; in this case, only library clones containing one or more copies of the target gene will yield amplification products. These clones can then be selected for sequencing. Alternatively, library clones can be chosen at random and end sequenced using primers matching the DNA insertion site. From these short snippets of sequence, clones of interest can be identified and shotgun sequenced to completion (see Chapter 3)—each clone provides a snapshot of one of the genomes present in the original DNA sample. The proteorhodopsin and ammonia oxidation examples described previously are two of a great many examples where such strategies have driven research forward in exciting new directions.

With next-generation sequencing technologies, it is even possible to assemble complete, or near complete, genomes from bulk community DNA. In this case, the entire metagenomic library is shotgun sequenced and assembled using standard bioinformatic tools. The complexity of the assembly problem can be significantly reduced by starting with an 'enrichment culture', in other words a culture established from the original environmental sample on a medium that favours the growth of only a handful of the original species. Enrichment strategies are often unsuccessful, however, and by definition such cultures do not represent the environment in its natural state.

An alternative strategy is to start with bulk DNA isolated from an environment that is naturally limited in terms of its microbial diversity. The feasibility of this latter approach was demonstrated in 2004 by a research team led by the Australian-born microbiologist Jillian Banfield. Here the target environment was a drainage site at the Iron Mountain Mine in Northern California. Acid mine drainages can occur naturally or as the result of industrial disturbances such as coal mining or construction; they are characterized by low pH (i.e. high acidity), high concentrations of toxic metals, and (not surprisingly) low levels of biodiversity. From biofilm samples taken hundreds of feet underground, the researchers extracted DNA and sequenced and assembled near-complete genomes from several bacteria and archaea. Reconstruction of their biochemical pathways provided insight into how the microbes are able to survive in such an extreme location.

Banfield's team has recently shown that with enough data and computational power it is possible to assemble complete genomes from complex microbial communities. This time focusing on sediment and groundwater samples taken near the Colorado River, the researchers generated ~4.6 *billion* Illumina sequence reads and were able to assemble ~2,500 complete or near-complete bacterial genomes and twenty-four archaeal genomes. The study led to the identification of no fewer than forty-seven new bacterial phyla and showed the remarkable extent to which the metabolisms of microbes in nature are intertwined. Metabolites generated by one organism within these communities were predicted to feed into the biochemical pathways of another and vice versa. Such observations go a long way to explaining why so much of microbial diversity is recalcitrant to culturing.

It should be noted that, strictly speaking, genomes derived from the deep sequencing of metagenomic libraries are artificial mosaics, in other words they are composites of sequence reads derived from multiple organisms. This is because no matter how low the level of microbial diversity, an environmental sample is a snapshot of a living community, not a pure culture—each individual DNA fragment comes from the genome of an individual cell, which was itself part of a larger population. Such mosaicism can, however, be highly informative. In the case of Banfield's acid mine drainage study, fine-scale analysis of metagenomic sequence variation allowed the researchers to infer patterns of gene exchange occurring within the populations of bacteria and archaea that comprise these acidic biofilms.

Metagenomics has thus far proven less useful as a tool for exploring the diversity of eukaryotic microbes. There are various reasons for this. Nuclear genomes typically have a much lower coding density than those of bacteria and archaea—eukaryotic genes are more spread out—and the genes themselves are usually interrupted by introns. Consequently, a randomly sequenced fragment of eukaryotic DNA is thus much less likely to be informative than a prokaryotic one. The large and repetitive nature of eukaryotic genomes also makes them more difficult to assemble than those of prokaryotes, and this is especially true when the sequence data are derived from a metagenomic library. Finally, far fewer eukaryotic genomes have been sequenced than prokaryotic ones, and without quality reference genomes available from an evolutionary diverse range of cultured organisms, the task of making sense of environmental sequence data becomes much more difficult. A handful of studies have used metagenomics to explore eukaryotic microbial diversity in nature, but in many respects the field is still in development.

Single-cell genomics and transcriptomics

A complementary approach to metagenomics is the application of DNA amplification technologies to the study of individual cells. By definition, 'single-cell genomics' (or single-cell sequencing) does not rely on the availability of a culture, and it sidesteps the problem of assembling sequence reads derived from bulk community DNA. The approach depends on the placement of a single cell in a reaction vessel, which can be achieved using various means. In the case of large protists (i.e. microbial eukaryotes excluding plants, animals, fungi, and their specific relatives), individual cells can be isolated by hand directly from an environmental sample using a fine-tipped capillary tube. More sophisticated methods can also be used, including use of a flow cytometer, an instrument capable of sorting cells in a high-throughput fashion based on their physical and chemical characteristics.

After cell sorting, a DNA amplification step must be carried out. This is necessary because the quantity of DNA inside even the largest eukaryotic cells is insufficient for sequencing. DNA is amplified using one of several molecular techniques, most commonly 'multiple displacement amplification'. This technique uses a special viral-derived DNA polymerase and a randomly generated mixture of primers to amplify the genome in a non-specific fashion. Once amplified, a next-generation DNA sequencing technique is used to generate the data, and the reads are assembled to produce a genome sequence (see Chapter 3). Single-cell genomics has proven effective for the study of prokaryotes and eukaryotes with small genomes, and combined with single-cell transcriptomics, makes it possible to determine not only the suite of genes contained by individual organisms within a population, but also how patterns of expression change from cell to cell, for example, in response to viral infection.

At the present time, one of the main limitations of single-cell 'omic' technologies is amplification bias. Carefully controlled experiments using microbes whose genomes have already been sequenced have shown that not all regions of a genome amplify to the same extent. This has led to concerns over reproducibility and uncertainties about the degree of completeness of single-cell-derived genomes and transcriptomes. The approach nevertheless can provide valuable data when it is not possible to obtain 'omic' information using traditional means. It has also proven useful for the study of the cells that make up macroscopic organisms. For example, human cancers can now be studied cell by cell, providing fresh insight into the mutations that arise as the disease progresses and how cancer cells change in response to the drugs being used to eradicate them (see Chapter 4).

The incredible expanding virosphere

In the same way that metagenomics has transformed our understanding of the diversity of cellular life, the use of culture-independent molecular tools has opened a Pandora's box of viral diversity. Marine viruses have been of particular interest, and with good reason: at ~10 million viral particles per millilitre (mL), they are the most abundant biological entities in the sea. According to the Canadian virologist Curtis Suttle, they carry out 10^{23} infections per second and contain as much carbon as seventy-five million blue whales. Stacked end-to-end, all the viruses in the ocean would extend beyond the nearest sixty galaxies.

But despite their astonishing abundance, viruses are easy to ignore. Prior to DNA extraction, environmental samples are typically size-fractionated so as to exclude cells that are larger or smaller than the organisms of interest, and because many (but not all) viruses are much smaller than prokaryotes, the viral fraction passes through the filters and is usually discarded. Even if one is specifically interested in viruses, their extreme genetic diversity can be problematic. Viruses don't have rRNA genes and there is in fact no single gene that is shared by *all* viruses that could serve as a universal molecular marker. This limits the usefulness of gene-specific environmental PCR approaches to the discovery of novel viruses. Shotgun metagenomic sequencing has proven ideal for the task.

The first viral metagenomic datasets were generated in the early 2000s, targeting marine sediments, hot springs, as well as the human gut and faeces. A common theme among all of these samples was the retrieval of a high percentage of genes of unknown function. Viral genomes are uniformly gene dense, but at least 50 per cent of the genes identified in metagenomic samples show no obvious sequence similarity to genes in other viruses or in prokaryotes or eukaryotes. In some cases this fraction exceeds 95 per cent. This is presumably a function of the fast evolutionary rates of viral genes, combined with the fact that so little of the 'virosphere' has been explored using molecular tools. Regardless of the reasons, this tremendous diversity underscores just how little we know about viral diversity in nature and how difficult it will be to understand what these mystery genes do for the viruses in which they reside.

Textbooks tell us that viruses are potent but limited in terms of their gene repertoires. Influenza genomes, for example, are

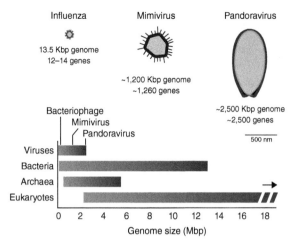

Influenza

13.5 Kbp genome
12–14 genes

Mimivirus

~1,200 Kbp genome
~1,260 genes

Pandoravirus

~2,500 Kbp genome
~2,500 genes

500 nm

Bacteriophage
Mimivirus
Pandoravirus

Viruses
Bacteria
Archaea
Eukaryotes

Genome size (Mbp)

0   2   4   6   8   10   12   14   16   18

18. Viral genome diversity relative to bacteria, archaea, and eukaryotes.

only ~13,500 bp long and have between twelve and fourteen protein-coding genes (see Figure 18), depending on the strain; the human immunodeficiency virus (HIV-1) has a genome of ~9,700 bp with nine genes. These are both RNA genomes. The genome of Hepatitis B is even smaller—3,020 bp and four genes—and is a mixture of single- and double-stranded DNA segments. But consideration of viruses beyond human health concerns paints a very different picture. Aided by next-generation sequencing, a wide range of DNA viruses infecting microbial eukaryotes, animals, and bacteria have been discovered. In some cases their genomes are so far beyond what is considered 'normal' that some have begun to revisit the age-old question of how viruses are related to cellular organisms.

In the early 2000s, a team of French researchers led by Jean-Michel Claverie and Didier Raoult characterized the genome of a virus dubbed Mimivirus. First isolated from a water-cooling tower in England during a pneumonia outbreak in 1992, the virus infects single-celled amoebae. Its name stands for 'mimicking microbe': at ~0.5 micrometres in diameter, Mimivirus was found hiding among similarly sized bacteria during filtration procedures. At ~1.2 million bp in size and with 1,262 protein-coding genes, the Mimivirus genome raised eyebrows. In 2013, an even larger viral genome was sequenced, that of the aptly named Pandoravirus, which also infects amoebae. The Pandoravirus genome is ~2.5 million bp and has ~2,500 genes. This is substantially larger than many prokaryotic genomes, and even those of some eukaryotes (see Figure 18). Only ~10 per cent of the genes in Mimivirus and Pandoravirus can be assigned a putative function based on sequence similarity alone.

One remarkable feature of the Mimivirus and Pandoravirus genomes is the presence of sequences that are similar to prokaryotic and eukaryotic translation proteins. This is unprecedented and perplexing, given that viruses do not have ribosomes and are thus incapable of making their own proteins. Another noteworthy discovery is that giant viruses can themselves have their own viruses. These 'virophages' have been shown to negatively impact the reproduction of the giant viruses (and thus benefit the host amoeba), putting a new biological twist on the phrase 'the enemy of my enemy is my friend'. Such observations, combined with the overlap in size between giant viral genomes and those of cellular organisms, have fuelled speculation that perhaps giant viruses belong to an ancient, previously unrecognized, fourth domain of life. If so—if giant viruses do indeed form an evolutionarily coherent entity—one would expect their genes to branch together in phylogenetic trees.

For the most part this is not the case. Of those genes in the Mimivirus and Pandoravirus genomes amenable to phylogenetic analysis, many are in fact most similar to those found in amoebae and bacteria. This is consistent with the hypothesis that their 'giant' genomes are the product of recent expansion; giant viruses could have evolved from viruses with smaller genomes by picking up genes from the eukaryotic cells that they infect, as well as from bacteria in their immediate environment. There is still debate over which of these two scenarios more closely reflects reality, but if one considers viruses as a whole, there is ample precedent for the 'viruses as gene pickpockets' idea. For example, the genomes of cyanophage—viruses that infect cyanobacteria—are well known to possess photosynthesis-related genes derived from their hosts (the cyanophage themselves do not photosynthesize) and to shuttle them between different strains of the bacteria. Do viruses belong on the tree of life? The jury is still out, but what is inescapable is that viruses are firmly 'plugged in' to the tree of life and capable of shuttling genes from branch to branch. As long as there have been cells on Earth there have been viruses infecting them.

The human microbiome

A 70 kilogram man 1.7 metres in height is comprised of ~30 trillion cells, ~25 trillion of which are red blood cells. That same man will have a total bacterial cell count of ~39 trillion—our cells are outnumbered ~1.3 to 1 by those of bacteria. The numbers of archaea and microbial eukaryotes living on and within us are not so easily quantified; although they clearly impact our health, bacteria outnumber them both by several orders of magnitude. These are extraordinary numbers, all the more so considering that they do not include the countless viruses that prey on our microbial companions. Our bodies are complex ecosystems and like any other natural environment, they are subject to perturbation.

Started in 2008, the US-based Human Microbiome Project (HMP) typifies the efforts of the broader international biomedical research community in trying to understand the roles of our resident microbes in health and disease. Efforts have focused primarily on five body sites: skin, oral, vaginal, gut, and nasal/ lung. Next-generation-enabled shotgun sequencing of microbial DNA extracted from these sites has been used to build upon

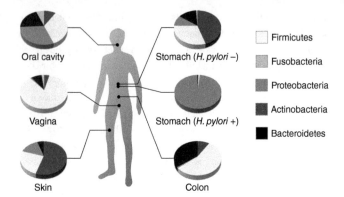

19. Relative abundance and diversity of bacteria on and in the human body.

earlier PCR-based rRNA gene-sequencing studies, which hinted at the astonishing diversity to be found in these and other regions of the body. For example, we now know that the skin microbiome is comprised of ~1,000 distinct bacterial species from perhaps as many as nineteen different phyla. This ~2-square-metre region of the body is dominated by members of the Actinobacteria (>50 per cent) and to a lesser extent, Firmicutes (~25 per cent) (see Figure 19). Although well known, the skin-associated firmicute bacterium *Staphylococcus* is not particularly abundant, comprising only ~5 per cent of the skin microbiome on the basis of rRNA amplicon analyses. (Note that because of HGT and pan-genomes (see Chapter 5), defining a prokaryotic 'species' can be problematic; in metagenomics, researchers typically define distinct species as organisms sharing <97 per cent rRNA gene sequence identity.)

One of the key take-home messages of the past decade of human microbiome research is just how variable community composition can be, both within and between healthy and diseased individuals. In contrast to skin, the gut microbiome is dominated by the presence of anaerobic bacteria, perhaps 500 species in total, thirty of which are particularly abundant. Some of these bacteria can produce vitamins such as K and H, and can improve the efficiency with which we digest plant-derived food. This is because they produce enzymes that break down certain carbohydrates into fatty acids, which we then digest using enzymes encoded by our own genes.

A recent metagenomic investigation showed that while Japanese individuals possess the gut bacterium *Bacteroides plebeius*, North Americans do not. This is significant because the genome of this particular species encodes an enzyme capable of degrading seaweed-derived polysaccharides. Curiously, other than *B. plebeius*, the gene appears to only exist in marine bacteria. It has thus been proposed that the Japanese *B. plebeius* acquired this 'useful' gene by HGT from marine bacteria, which were ingested along with the raw seaweed used to make sushi.

The proteobacterium *Helicobacter pylori* is another example of a patchily distributed microbe. Although this organism has been linked to ulcers and stomach cancer, >80 per cent of those who harbour it do not exhibit symptoms. Scientists are thus trying to understand the extent to which *H. pylori* plays a role in the 'normal' ecology of the gut. What is clear is that *H. pylori* is highly adapted to life in this acidic environment and upon infection, the levels of Firmicutes and Actinobacteria in the gut drop significantly (see Figure 19). This change can be permanent. An increased ratio of skin-associated Firmicutes to Actinobacteria has been associated with psoriasis, and links between obesity and a reduced Bacteroidetes-to-Firmicutes ratio have also been suggested.

One way that microbiome researchers have attempted to control for human genetic variability is by studying twins. Interestingly, while the microbiomes of monozygotic twins are more similar to one another than those of unrelated people, the differences are not dramatic. The picture emerging from bioinformatics-based metabolic profiling of metagenomic data is that function is more

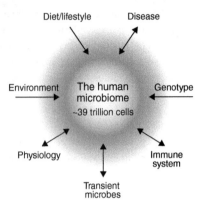

Diet/lifestyle   Disease

Environment   **The human microbiome**   Genotype
                        ~39 trillion cells

Physiology   Immune
                                                 system

                      Transient
                      microbes

20. Internal and external factors that influence the composition and relative abundance of microbes in the human microbiome.

important that taxonomy. That is, the overall metabolic characteristics of a particular microbial community are more stable from person to person than the underlying species composition. It is now clear that a complex array of factors contribute to the establishment and maintenance of the human microbiome over the course of a lifetime (see Figure 20). The ultimate goal of research in this area is to establish a reference point for what is a 'normal' microbiome for a healthy individual. To reach that goal we need to better understand how much—and why—the microbiome varies; how its composition changes as we age; and how much microbiome variation is influenced by differences in our genotype and physiology. Human microbiome research is thus increasingly seen as an important arm of medical genetics.

Bioprospecting in the metagenomic era

The World Health Organization defines bioprospecting as 'the systematic search for and development of new sources of chemical compounds, genes, microorganisms, macroorganisms, and other valuable products from nature'. Metagenomics has rapidly become a core technology in this competitive area of applied research. At its best, bioprospecting combines bioinformatics with high-throughput experimental strategies for detecting novel biochemical activities in microbial communities. The latter allows researchers to bypass the challenge of accurately predicting gene functions from sequences alone. Indeed, one initially need not have any sequence data; all that matters is that environmental DNA fragment X is associated with activity Y. The gene (or genes) underlying the activity can be sequenced later.

The most widely used method for function-based screening is the use of so-called 'expression' vectors. Here environmental DNA fragments are cloned into vectors containing sequences that, when inserted into a particular host organism, will trigger the host to express the gene(s) located on each fragment. The bacterium *E. coli* is a commonly used host in such experiments, but there are other options, including the single-celled yeast *Saccharomyces*. One then needs a way to sift through all of the different bacteria (and their snippets of environmental DNA) for the presence of the activity/protein/gene of interest. Here the 'readout' is often detection of a particular metabolite, which can, for example, make its presence known by causing the growth medium to change colour. With a sufficiently robust assay, the protein products of many *E. coli* can be screened at the same time. Parallelization is critical: there are many technical challenges associated with expression vector-based screening and it is in the researcher's best interests to maximize the chances of getting a 'hit'.

The potential for discovery using such an approach is significant. For example, functional screening of DNAs extracted directly from soil—one of the most diverse microbial communities known to science—has resulted in the identification of many novel prokaryotic proteins with important biotechnological applications. These include fat-digesting enzymes used in the food, detergent, and cosmetics industries, as well as enzymes with anti-cancer and anti-microbial properties. Marine fungi have yielded enzymes that are naturally resistant to extreme temperatures and pH. In principle, any environment in which microbes thrive is a potential source of novelty. A recent metagenomic survey of human-associated bacteria led to the intriguing discovery of genes linked to the production of the antibiotic lactocillin. Subsequent experiments showed that lactocillin is indeed produced by microbes inhabiting the vagina, presumably as a natural protection against pathogens. The human microbiome may prove to be a rich—and profitable—source of drugs with anti-microbial properties.

The genomic encyclopedia

In the past, financial and technical constraints often meant that organisms chosen for genome sequencing were the easiest to grow and of biomedical and/or biotechnological relevance. The Genomic Encyclopedia of Bacteria and Archaea (GEBA) is an ambitious, international project that aims to overcome the sampling biases that dominated the first fifteen years of microbial genome sequencing. GEBA researchers, led by the US Joint Genome Institute's Nikos Kyrpides, point out that only ~30 per cent of bacterial and archaeal 'type strains'—named prokaryotic species with a cultured representative—have had (or are having) their genomes sequenced. GEBA aims to fill in the gaps by sequencing the genomes of prokaryotes based on their taxonomic affiliation, not short-term practical relevance.

The overarching goal of GEBA is 'to sequence the genome of at least one representative of every bacterial and archaeal species whose name has been validly published'—all 11,000 of them. As ambitious as it sounds, it is increasingly realistic given rapidly evolving sequencing technologies and bioinformatics capabilities. Given that hundreds of new bacterial species are named each year, the finish line for such an effort is a moving target. Regardless, a GEBA-style approach to documenting prokaryotic diversity will make lasting contributions to both basic and applied research.

Of particular significance will be GEBA's role in providing a comprehensive reference point for the interpretation of metagenomic data, allowing researchers to make better sense of the gene sets found in diverse organisms living in diverse habitats.

There is currently no equivalent to GEBA for microbial eukaryotes but there should be. Consider that at the present time, more fruit fly genomes have been sequenced than for all of the Rhizaria, an exceptionally diverse group of eukaryotes that goes back more than a half-billion years in the fossil record. Noteworthy rhizarians include foraminiferans, single-celled eukaryotes upon which the oil industry heavily depends for dating sediments; a wide range of amoebae known to be important players in soil ecology; and the radiolarians, abundant marine protists with mineralized skeletons, made famous by the drawings of Ernst Haeckel at the turn of the last century. Rhizaria are also related to stramenopiles (which include the diatoms in the sea that produce ~50 per cent of the world's oxygen) and alveolates, which are an eclectic mix of parasites and free-living organisms (see Figure 16). Rhizaria are just one under-studied protist group among many; culture collections are full of photosynthetic and non-photosynthetic protists, many of which have never had their ribosomal RNA genes sequenced let alone their genomes.

Microbial eukaryotes have not been entirely neglected, far from it. Building off the first yeast genome published in 1996, fungal comparative genomics has become a very mature field with hundreds of sequences completed, and the biomedical sciences have benefited greatly from the many dozens of genomes sequenced for pathogens such as the malaria parasite *Plasmodium* and *Trypanosoma*, the causative agent of African sleeping sickness and Chagas disease. Microalgal genomes are also an area of particular emphasis. As genome sequencing and assembly methods continue to improve, we should eventually see a systematic, taxonomically driven quest to explore the genomic diversity of microbial eukaryotes. It has been estimated that the genetic and cell biological diversity encompassed by protists is as high or higher than all of the animal and plant kingdoms combined. A concerted effort will be needed to fill in these huge knowledge gaps. The pay-off will be a much stronger framework for interpreting the wealth of metagenomic and single-cell genomic data coming off sequencers in labs around the world.

Chapter 7
The future of genomics

'Today's fiction becomes tomorrow's reality'. This oft-said phrase readily applies to molecular biology and genomics. In this final chapter, we explore some of the ways in which genome science will continue to expand the frontiers of human knowledge, as well as change our world and the way we interact with it. Important ethical, legal, and social issues have already been raised. Who owns your genome sequence and who has the right to access the information it contains? Should it be legal to make 'designer babies' by editing the DNA of human embryos? Should we bring extinct species back to life and introduce them into the wild? As genomics-enabled personalized and reproductive medicine become increasingly commonplace, such questions will no longer be hypothetical.

Genomes on demand—and in three dimensions

By the time you read this book, some of the information it contains will already be out of date. Although it is impossible to predict exactly where genome sequencing will go and how it will get there, three general directions are obvious: the DNA sequencers of the future will be smaller, faster, and more accurate. In terms of size, the future is already here. Oxford Nanopore Technologies (ONT), the company that developed the pocket-sized MinION (see Figure 5), recently announced the SmidgION, a thumb-sized sequencer that will plug into your smartphone. Precisely what this instrument will be capable of is not yet clear, but if the rapid maturation of MinION sequencing is any indication, it could be a game-changing development.

Portability is one thing, but what is missing is the ability to prepare DNA samples on the fly. ONT is working on that too. New in-the-field tools and reagents are being developed that will make it possible to go from environmental sample to genome sequencing in minutes. The possibilities are endless. Imagine tracking an infectious-disease outbreak in real time; or monitoring a hospital ward on-site for the spread of antibiotic-resistant bacteria; or evaluating the progress of genetically engineered bacteria at a wastewater-treatment plant.

Back in the lab, genome sequencing will get quicker and quicker. DNA sequence reads will continue to get longer until it becomes routine to read from one end of a chromosome to the other. The genome assembly problem will cease to exist. The genomes and transcriptomes of single cells taken from anywhere and everywhere will be sequenced with high accuracy. The possibilities are infinite here as well. It will be possible to sequence RNA directly without having to first convert it back to DNA (the MinION is in fact already capable of this). Protein sequencing will make a comeback; in principle, any polymeric substance that can be threaded through a nanopore can be sequenced, including the amino acid chains that comprise our proteins. We may eventually abandon nanopores in favour of high-powered electron microscopes trained to sequence individual molecules at atomic resolution. The seeds of this technology are already being sown.

We will have a comprehensive understanding of the genome in three dimensions. Over the past decade, research at the interface of genomics and microscopy has provided deep insight into the biology of chromosomes in living cells, but genomicists (myself included) still have a tendency to think of genomes as sterile, linear entities. Nothing could be further from the truth. Understanding the nature of long-range interactions between distinct chromosomal regions is key to understanding how genes are switched on and off, in normal cells and in cancerous ones. It is also critical to figuring out the nuts and bolts of fundamental processes such as cell division and genetic recombination. In the cell biology of the future, the genome will come alive.

And we will completely sequence 'the' human genome—every last base pair of it. This will be achieved by combining the latest and greatest high-throughput methods with old-school thinking about 'physical maps', a marriage of the competing strategies that defined the initial race for the human genome. Your genome will be sequenced and so will mine, if we so choose. The biggest challenge facing genomics in the future will be what to do with the data.

DNA as storage device

In 2011, Jonathan Rothberg needed a prominent test subject for his new Ion Torrent sequencer. He asked Intel co-founder Gordon Moore, a fitting choice on various levels. 'Moore's Law' refers to the fact that every eighteen months or so, the number of transistors that can be packed onto an integrated circuit doubles. This has been going on since the mid-1960s. Unfortunately, this steady increase in electronics performance is nearing an end. With current silicon-based technologies, the laws of physics place an upper limit on transistor density. Potential solutions to the problem, such as the development of quantum computers, are still firmly in the realm of science fiction. What about DNA? Might it be possible to convert the bits and bytes of our digital world into As, Cs, Gs, and Ts for the purposes of data storage?

It has already been done. In 2012, Harvard genomics pioneer and futurist-provocateur George Church published a book, *Regenesis*, in which he and co-author Ed Regis explore life in the age of genetic engineering. At the back of the book Church describes how he converted a digital version of its content (53,426 words and eleven images) into the language of DNA. All the 1s and 0s were recoded and synthesized as many thousands of short (159-nucleotide) strings of DNA bound to microchips. The 'book' was then amplified by PCR and 'read' on an Illumina DNA sequencer. Ten errors were detected.

Using a different recoding strategy, Nick Goldman and colleagues at the European Bioinformatics Institute in the UK have used DNA to store images, Shakespearian sonnets, and Watson and Crick's famous double-helix paper. Church, Goldman, and colleagues are not the first to store data in DNA—this was achieved in 1988—but they are prominent among an ambitious group of scientists pushing the boundaries of what is possible. Progress is rapid. Using a storage strategy called DNA Fountain, Yaniv Erlich and Dina Zielinski at the New York Genome Center recently encoded an entire computer operating system and a movie (~2 megabytes of compressed data) in DNA at a density of 215 petabytes per gram of DNA. It was decoded with perfect accuracy by Illumina sequencing.

All this is impressive, but there are significant challenges to overcome before DNA-based hard drives could become mainstream. The biggest one is DNA synthesis, which is currently much too expensive and slow. It cost US$7,000 to synthesize the 72,000 DNA fragments used in the Erlich–Zielinski experiment (with another US$2,000 being spent on sequencing). And even if synthesis was cheap, the fastest DNA synthesizers cannot produce the molecules fast enough to be of any use for data storage on a commercial scale.

There is also the issue of data retrieval. If your data files are stored in DNA, how do you access only the files you want, as you can on your computer, without having to retrieve (i.e. sequence) all of them at the same time? One approach that has already been successfully employed is to tag each of the files with a unique barcode during DNA synthesis. Using PCR primers designed to specific tags, one can amplify and sequence only the files you want. PCR takes time, however, and is error-prone. At present, DNA shows more promise as a medium for long-term data archiving than short-term storage. Whatever approach is used for writing and reading DNA, it needs to be able to detect and correct errors introduced by nature's imperfect enzymes.

The era of digital DNA (if there ends up being one) is still a long way off. Million-fold improvements in the scale and speed of DNA synthesis are needed if it is to become a viable solution to the world's growing data storage problem. But it is worth considering that the cost of DNA sequencing has dropped two-million-fold since 2003, a pace that exceeds Moore's Law. And consider this: at a theoretical maximum density of 2 bits per nucleotide, *all* of the world's data could be stored in a mere kilogram of DNA. That's a lot of DNA—it's also a lot of data! It is not difficult to envision a future in which the world's data archiving centres are full of DNA chips instead of computer hard drives.

Designer genomes—rewriting the book of life

An appealing aspect of DNA-based data storage is that the tools are all around us. Appropriately stored, the data will be accessible to future generations using the information processing enzymes of biological systems (and even if the DNA becomes degraded, ancient DNA research has shown that severely damaged DNA can still be read). The field of biotechnology was founded upon nature's bounty. The enzymes first used in the 1970s to cut and splice together DNA from different organisms are derived from bacteria; their 'day job' is to destroy incoming viral DNAs. The PCR amplification enzyme 'Taq' polymerase was originally isolated from *Thermus aquaticus*, a bacterium that lives at 70°C in the hot springs of Yellowstone National Park. As we learned in Chapter 6, metagenomics has become a powerful tool for harnessing the biochemical potential of the natural world.

One of the most exciting recent developments in biomedical research goes by the name of CRISPR, which stands for 'clustered regularly interspaced short palindromic repeats'. CRISPRs exist naturally in the genomes of bacteria and archaea, where they form part of what has been described as a prokaryotic immune system. The elucidation of the biology of CRISPRs and CRISPR-associated (Cas) proteins led to the development of a powerful DNA editing technology by Emmanuelle Charpentier, Jennifer Doudna, and Feng Zhang. CRISPR is in fact one of several DNA editing methods that have been developed in recent years, but it is faster, more flexible, and more accurate than the others; it has thus become the tool of choice. (Unsurprisingly, there has been a fierce legal battle for the right to patent CRISPR technology.)

In prokaryotes, CRISPRs and Cas proteins evolved in concert to degrade foreign DNAs entering the cell. This is achieved in part by the production of small, CRISPR-derived RNAs that help target the incoming DNA for destruction by Cas enzymes. In the lab, the tag-team action of CRISPR-Cas9 has been exploited to edit a genome at will (see Figure 21). A synthetic 'guide' RNA (gRNA) is first made with a sequence that matches the specific site in the genome that one wishes to edit (the target site). The gRNA-associated Cas9 enzyme then cuts the target site on both DNA strands. Finally, user-provided 'donor' DNA is inserted into the cut site by the cell's natural DNA repair proteins. Any DNA of interest can be inserted into virtually any location in the genome.

Although still quite new, the CRISPR-Cas9 editing system has already been used in a wide variety of basic and applied research applications. There has been both fanfare and controversy. For example, it has been used to repair disease-causing mutations in mice. This is achieved by injecting the desired CRISPR-Cas9 components into an early-stage embryo or zygote (fertilized egg).

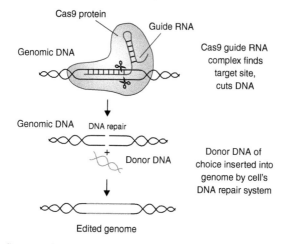

21. Genome editing using CRISPR-Cas9.

Once the DNA has been edited, the change is permanent; the organism develops with the modified sequence in its genome, and passes it on to future generations. In the lab, the genomes of human cells have also been edited using CRISPR-Cas9. One of the key challenges and concerns is the issue of off-target modifications. The CRISPR-Cas9 system is very good at making the desired target-site edits but unintended changes elsewhere in the genome can also occur. This is undesirable and, depending on the application, potentially dangerous. Researchers are exploring various ways to minimize the problem and, hopefully, eliminate it. All things considered, CRISPR-Cas9 shows considerable promise for gene therapy and stem cell-related therapeutics. For safety and ethical reasons, however, there is concern over the use of CRISPR-Cas9 for editing human zygotes.

Synthetic biology

At the interface of biology and engineering is a rapidly advancing discipline called synthetic biology. It takes a reductionist, rational, and modular approach to the design, modification, and even *de novo* assembly of cells and cell parts. As one might expect, synthetic biology is built upon several decades of developments in molecular biology and takes advantage of the latest tools such as CRISPR-Cas9. However, with inspiration from classical engineering, the field has evolved a language of its own. The fundamental units of synthetic biology are so-called 'BioBrick parts', DNA sequences containing functional elements such as protein-coding genes and the promoters that regulate their expression. These parts are delineated by defined enzyme-digestible regions, which allow researchers to insert them into bacterial cloning vectors and to mix and match them in order to build 'devices'. Devices are sets of BioBrick parts that have been put together to carry out a particular sub-cellular function. Devices are then combined into 'systems' that can perform larger sets of cell biological tasks.

Many (but not all) areas of synthetic biology place significant emphasis on standardization and 'open-source' science. This focus is embodied by the international Genetically Engineered Machines (iGEM) competition, held every year in Boston. The event has grown over the past decade to include participants from around the world, including high-school and university students. The goals of iGEM are to raise awareness of, and interest in, synthetic biology and, at the same time, generate standardized BioBrick parts for the benefit of the field as a whole. Against this community oriented backdrop is applied research carried out in industrial settings, often in collaboration with scientists in academic laboratories.

One of the most exciting uses of synthetic biology is to engineer cells as 'factories' for the production of biological compounds of use to science and society. In principle this can be achieved using two different strategies, 'bottom-up' and 'top-down'. The latter approach starts with natural cells and seeks to modify their genetic make-up in a stepwise fashion. A landmark achievement was the development of a system for producing the anti-malarial drug artemisinin. The compound is a terpenoid that exists naturally in *Artemisia annua*, a plant used in traditional Chinese medicine; it can be extracted directly from the source but only with great difficulty and expense. Synthetic biologist Jay Keasling and colleagues re-engineered the metabolic pathways of *E. coli* and, later, yeast to produce artemisinin in the lab. Other recent examples of 'metabolic engineering' include use of yeast for the production of isobutanol as a biofuel; *E. coli* for the synthesis of chemical precursors used in the production of spandex; and *Corynebacterium* for the commercial production of amino acids such as lysine.

Top-down metabolic engineering has enormous potential but it is also hard work. The complexity of living cells is such that it can take five to ten years and many millions of dollars to bring a new 'cell factory' to the point where it can be used to efficiently and economically produce a chemical compound of interest (artemisinin took ten years and ~US$50 million). It is not enough to simply insert the right genes into the genome of a workhorse organism; Keasling's work underscores the importance of the 'design-build-test-learn' cycle so familiar to engineers. In the case of metabolic engineering, the cycle often needs to be repeated numerous times in order to achieve the desired output. But the rate of progress is accelerating, aided by a growing synthetic biology toolkit and knowledge gained by trial and error.

Bottom-up approaches to synthetic biology are equally challenging. Ideally we could build up a cell piece by piece with a specific outcome in mind, unconstrained by the inherent 'messiness' of life. Having evolved by natural selection, even the simplest of cells have complex and often redundant genetic circuitries that may or may not be relevant to the purposes of the researcher but nevertheless must be taken into account. Craig Venter and colleagues have combined top-down and bottom-up approaches in their research on what is perhaps the 'simplest' organism known, a bacterium called *Mycoplasma*. In 1999, they used gene knockout technology to predict a minimal set of ~375 genes that the organism needs for life. In 2016, Venter and company synthesized a rearranged version of the organism's genome from scratch in the lab, and successfully used it to 'boot up' a *Mycoplasma* cell whose own genome had been removed.

But even *Mycoplasma* is a complex organism with hundreds of genes. True bottom-up approaches to synthetic biology are more akin to organic chemistry. They focus on cell-free systems in which synthetic molecules such as RNAs are tinkered with in order to elucidate, and ultimately control, fundamental biochemical processes. It remains to be seen whether cells will ever be built from the ground up, but we will no doubt learn a lot by trying. The spirit of this arm of synthetic biology is perhaps best captured by the famous blackboard quote of Richard Feynman: 'What I cannot create, I do not understand'.

Resurrection genomics

'De-extinction' (or resurrection) biology is a rapidly advancing and controversial area of science that explores the possibility that extinct species might one day be brought back to life. Biologists presently recognize three general approaches that could conceivably be used to make this a reality: back-breeding, cloning, and genome editing. Strictly speaking, back-breeding does not result in the resurrection of an extinct species; rather, it employs the selective breeding of present-day organisms in order to bring back specific physical and/or behavioural characteristics that have been lost over time.

The best-known example of back-breeding involves the aurochs, the extinct wild cattle species from which modern domesticated cattle emerged. By systematically mating particular breeds of cattle exhibiting aurochs-like traits, the goal is to recreate the sharp-horned (and bad-tempered) wild creatures that disappeared from Europe in the early 1600s. Attempts to bring back the aurochs date back to the 1920s when the German zoologists Lutz and Heinz Heck carried out a back-breeding programme that resulted in a breed known as Heck cattle, which resembles the aurochs in some (but not all) of its characteristics. (The Heck brothers also bred into existence the so-called Heck horse, modelled after the extinct Eurasian wild horse, or tarpan.) Present efforts to back-breed the aurochs take advantage of genetic and genomic data from phenotypically diverse breeds of cattle as well as an actual aurochs genome, sequenced from DNA extracted from a ~6,800-year-old humerus bone found in Britain.

De-extinction by cloning relies upon somatic cell nuclear transfer, the procedure famously used to clone Dolly the sheep in the 1990s. It begins by implanting the nucleus of a somatic (body) cell taken from an animal into an egg cell that has been stripped of its own nucleus. Stimulated by an electric shock, this hybrid cell will then begin to divide, and the resulting embryo can be implanted into a host mother and brought to term. Straightforward though it seems, the procedure is technically demanding and generates viable offspring at a very low frequency. And in order to use cloning for de-extinction, researchers must have access to cells of the extinct organism of interest (otherwise there is no nucleus to transfer); this is a rare situation, possible only in cases where a species has gone extinct very recently and scientists have cultured and/or frozen materials. While interspecies nuclear transfer has been demonstrated in the lab between cells of the endangered gaur bull and domestic cows, it seems likely that the greater the evolutionary distance between the two organisms, the less likely the chance of success due to incompatibilities between the donor nucleus and the recipient cell. The long-term prospects of this approach to de-extinction are at present unclear.

Finally, CRISPR-Cas9 (or some other editing tool) may be used to engineer the genome of a living organism to resemble that of an extinct one, in whole or in part. This approach relies on the availability of genome sequence data from the extinct species of interest, which places constraints on the age of the organism one wishes to resurrect. Reliably extracting DNA from material more than a few thousand years old is extremely difficult (see Chapter 5), and the 'shelf-life' of ancient DNA depends greatly on the conditions of preservation. For now at least, large-scale genome editing is also painstaking work, with many variables to account for before cells engineered in the test tube might be used for inter-species nuclear transfer.

For these and other reasons, de-extinction by genome editing is presently still science fiction—but only just. George Church has announced that he and his research team may be in a position to bring back the woolly mammoth as early as 2019. Research indicates that mammoth populations took a serious hit between 15,000 and 10,000 years ago; the very last of their kind appear to have died out ~5,500 years before present on St Paul Island off the west coast of what is now mainland Alaska, and a mere 4,000 years ago on Wrangel Island north-east of Siberia. It is unclear what circumstances led to the demise of this iconic ice-age creature. One hypothesis is that we humans hunted the mammoth to extinction, but warming temperatures associated with the end of the last ice age likely also played a significant role.

Regardless, complete woolly mammoth genome sequences have been determined using DNA extracted from bones frozen into the tundra, and these sequences are being compared to the genome of the Asian elephant, the mammoth's closest known living relative. Church's team is systematically editing the genome of elephant embryos, endowing them with several dozen mammoth genes, including those predicted to play a role in hair and fat deposition, as well as those related to ear size. The result would be a hybrid elephant-mammoth embryo, which could conceivably be implanted and brought to term in an elephant or grown in an artificial womb.

Why might we want to resurrect the woolly mammoth, or at least a mammoth-like elephant better able to thrive in cold temperatures? It has been proposed that the return of such a creature would help to mitigate the effects of climate change by transforming the arctic tundra back to the so-called mammoth steppe grasslands, as it was during the last ice age when mammoths and other large grazers such as bison dominated the landscape. By trampling trees in summer and making holes in the snow in winter, the argument goes, mammoths would slow the rate at which the underlying permafrost is melting, which in turn would slow the release of greenhouse gases into the atmosphere. This provocative idea stems from the work of Russian scientist Sergey Zimov. Together with his son Nikita, Zimov has created Pleistocene Park, a refuge in the remote Siberian tundra, and has populated it with musk oxen, bison, and 're-wilded' horses. It is conceivable that a genetically engineered woolly mammoth look alike will find its place there in the not too distant future.

Various other creatures for which the possibility of de-extinction has been raised include the dodo, the Tasmanian tiger, and the passenger pigeon. The reasons differ from organism to organism, as do the concerns raised against them. At present it is unclear where all of this will lead. UC Santa Cruz evolutionary biologist Beth Shapiro has emphasized that de-extinction technologies need not be limited to the resurrection of extinct species. Indeed, the approach holds great promise in the area of conservation biology, for example, in cases where endangered species are low in number and in possession of insufficient genetic diversity to have a good chance of survival. This situation applies to the Asian elephant, which for various reasons (including the ivory trade) has seen its numbers in the wild decline from just under a million in the 1980s to ~40,000 today. Should we use genome science to do something about it?

The future of genomics

We return to where we began: the human genome and how knowledge of it will change our lives. With increasingly widespread personal genome sequencing, the age of personalized medicine is almost upon us. In some ways it is already here if we consider that genomic technologies are now commonly applied in diagnosing and treating diseases such as cancer (see Chapter 4). With genomics as a foundation, other 'omics' tools are finding their place in the personalized medicine toolbox as well. For example, transcriptomics is now being used to monitor the activity of genes whose expression levels are known to fluctuate in specific types of breast and prostate cancer. Therapeutic strategies can be tailored accordingly.

The so-called 'epigenome' is also an area of increased focus. Epigenetic changes are stable changes in gene expression that are not the result of altered DNA sequences. The methylation of cytosine residues in DNA is a particularly important epigenetic modification. The methylation level of genes has proven to be diagnostic for particular drug sensitivities, including chemotherapeutic agents; having this knowledge is thus useful when making decisions about treatment regimes.

In the future, personal genomics will be used not only for the diagnosis and treatment of disease, but for the purposes of disease prevention. It will also become a useful tool for those wishing to make informed lifestyle decisions so as to maximize their long-term well-being. This includes managing one's diet and exercise, use of over-the-counter medications, and adverse reactions to prescription drugs. In short, genome sequencing will be for healthy people too.

However, it is important to remember that genome sequence interpretation is not an exact science, not even close. A genome cannot tell us everything or even most of what we might want to know about our future health and disease susceptibilities. Genome sequences vary greatly from person to person, and we all carry mutations that have in some way been linked to disease. In some instances, like the breast cancer-associated genes *BRCA1* and *BRCA2*, the link to disease is strong (women with specific mutations in these genes have an 80 per cent chance of developing breast or ovarian cancer). In most cases, however, the risks are much less clear and dependent on a myriad of factors, only some of which are genetic.

Finally, among the biggest challenges facing personalized medicine is the issue of privacy. Physicians will increasingly encounter patients who have had their genomes completely sequenced. Their reasons for having done so will vary. So too will the extent to which they have educated themselves on what their sequence can and cannot tell them, and what the implications might be if significant disease risks are discovered. Who has a right to know and how should privacy be maintained? There is a balance to be struck between maximizing the utility of genomic data in disease prevention, diagnosis, and treatment on the one hand, and minimizing the potential for discrimination on the other. This includes, but is not limited to, discrimination in employment, access to insurance and health care, and various social contexts.

Genomics will change our world in ways we can and cannot imagine. In CRISPR-Cas9, we have a technology that can cure disease and enable us to permanently modify the human germline. The social, legal, and ethical implications are significant. Resurrection genomics and certain aspects of synthetic biology raise thorny ethical issues of their own. Both have the potential to change the course of evolution on our planet. A common concern raised by both scientists and non-scientists relates to the extraordinary pace of technology development in the molecular sciences. All too often it seems as though there has not been sufficient time to explore the implications of a new technology before it has already been used. Others argue that time is short; we need to be aggressive in our efforts to use all the tools available to us to solve the looming energy crisis, address human-induced climate change, and deal with a variety of other societal concerns that could conceivably be tackled using the tools of synthetic biology and genomics. Deciding the 'best' paths forward will not be easy but one thing is clear: there will be no shortage of creative, ambitious, and forward-thinking scientists willing to push the boundaries of what is possible. It is up to each of us to educate ourselves so that we can have an informed opinion on what is prudent.

References

Chapter 2: How to read the book of life

S. Goodwin, J.D. McPherson, and W.R. McCombie (2016). Coming of age: ten years of next-generation sequencing technologies. *Nature Reviews Genetics*, 17: 333–51.

J.M. Heather and B. Chain (2016). The sequence of sequencers: the history of DNA sequencing. *Genomics*, 1–7: 1–8.

J. Quick et al. (2016). Real-time, portable genome sequencing for Ebola surveillance. *Nature*, 530: 228–32.

J.M. Rothberg et al. (2011). An integrated semiconductor device enabling non-optical genome sequencing. *Nature*, 475: 348–52.

Chapter 3: Making sense of genes and genomes

R. Ekblom and J.B.W. Wolf (2014). A field guide to whole-genome sequencing, assembly and annotation. *Evolutionary Applications*, 7: 1026–42.

D. Sims et al. (2014). Sequencing depth and coverage: key considerations in genomic analyses. *Nature Reviews Genetics*, 15: 121–32.

H. Stefansson et al. (2005). A common inversion under selection in Europeans. *Nature Genetics*, 37: 129–37.

Z. Wang, M. Gerstein, and M. Snyder (2009). RNA-Seq: a revolutionary tool for transcriptomics. *Nature Reviews Genetics*, 10: 57–63.

Chapter 4: The human genome in biology and medicine

M.J.P. Chaisson et al. (2015). Resolving the complexity of the human genome using single-molecule sequencing. *Nature*, 517: 608–11.

R. Cordaux and M.A. Batzer (2009). The impact of retrotransposons on human genome evolution. *Nature Reviews Genetics*, 10: 691–703.

S.R. Eddy (2012). The c-value paradox, junk DNA and ENCODE. *Current Biology*, 22: R898–R899.

M. Gorski et al. (2017). 1000 genomes-based meta-analysis identifies 10 novel loci for kidney function. *Scientific Reports*, 7: 45050.

International Human Genome Sequencing Consortium (2001). Initial sequencing and analysis of the human genome. *Nature*, 409: 860–921.

S. Levy et al. (2007). The diploid genome sequence of an individual human. *PLoS Biology*, 5: e254.

S. Scherer (2008). *A short guide to the human genome.* Cold Spring Harbor Laboratory Press.

A. Telenti et al. (2016). Deep sequencing of 10,000 human genomes. *Proceedings of the National Academy of Sciences, USA*, 113: 11901–6.

J.C. Venter et al. (2001). The sequence of the human genome. *Science*, 291: 1304–51.

B. Vogelstein et al. (2013). Cancer genomic landscapes. *Science*, 339: 1546–58.

D.A. Wheeler et al. (2008). The complete genome of an individual by massively parallel DNA sequencing. *Nature*, 452: 872–7.

The 1000 Genomes Project Consortium (2015). A global reference for human genetic variation. *Nature*, 526: 68–74.

Chapter 5: Evolutionary genomics

F.H.C. Crick (1958). On protein synthesis. *Symposia of the Society for Experimental Biology*, 12: 138–63.

J. Dabney, M. Meyer, and S. Pääbo (2013). Ancient DNA damage. *Cold Spring Harbor Perspectives in Biology*, 5:a012567.

S. Fan et al. (2016). Going global by adapting local: a review of recent human adaptation. *Science*, 354: 54–9.

R.E. Green et al. (2010). A draft sequence of the Neanderthal genome. *Science*, 328: 710–22.

P. Holland (2011). *The animal kingdom: a very short introduction.* Oxford University Press.

A. Karnkowska et al. (2016). A eukaryote without a mitochondrial organelle. *Current Biology,* 26: 1274–84.

S. Kumar (2005). Molecular clocks: four decades of evolution. *Nature Reviews Genetics,* 6: 654–62.

I. Mathieson et al. (2015). Genome-wide patterns of selection in 230 ancient Eurasians. *Nature,* 528: 499–503.

J.O. McInerney (2014). The hybrid nature of the Eukaryota and a consilient view of life on Earth. *Nature Reviews Microbiology,* 12: 449–55.

M. Meyer et al. (2012). A high-coverage genome sequence from an archaic denisovan individual. *Science,* 338: 222–6.

R. Nielsen et al. (2017). Tracing the peopling of the world through genomics. *Nature,* 541: 302–10.

A. Sebé-Pedrós, B.M. Degnan, and I. Ruiz-Trillo. (2017). The origin of Metazoa: a unicellular perspective. *Nature Reviews Genetics,* 18: 498–512.

M. Slatkin and F. Racimo (2016). Ancient DNA and human history. *Proceedings of the National Academy of Science, USA,* 113: 6380–7.

S.M. Soucy, J. Huang, and J.P. Gogarten (2015). Horizontal gene transfer: building the web of life. *Nature Reviews Genetics,* 16: 472–82.

K. Zaremba-Niedzwiedzka et al. (2017). Asgard archaea illuminate the origin of eukaryotic cellular complexity. *Nature,* 541: 353–8.

Chapter 6: Genomics and the microbial world

K. Anatharaman et al. (2016). Thousands of microbial genomes shed light on interconnected biogeochemical processes in an aquifer system. *Nature Communications,* 7: 1329.

P. Bork et al. (2015). Tara Oceans studies plankton at planetary scale. *Science,* 348: 873.

I. Cho and M.J. Blaser (2012). The human microbiome: at the interface of health and disease. *Nature Reviews Microbiology,* 13: 260–70.

M.G. Fischer (2016). Giant viruses come of age. *Current Opinion in Microbiology,* 31: 50–7.

C. Gawad, W. Koh, and S.R. Quake (2016). Single-cell genome sequencing: current state of the science. *Nature Reviews Genetics,* 17: 175–88.

J.A. Gilbert and C.L. Dupont (2011). Microbial metagenomics: beyond the genome. *Annual Reviews of Marine Science,* 3: 347–71.

P. Hugenholtz (2002). Exploring prokaryotic diversity in the genomic era. *Genome Biology*, 3: R0003.1–R0003.8.

The Human Microbiome Project Consortium (2012). Structure, function and diversity of the healthy human microbiome. *Nature*, 486: 207–14.

N.C. Kyrpides et al. (2014). Genomic encyclopedia of bacteria and archaea: sequencing a myriad of type strains. *PLoS Biology*, 12: e1001920.

Chapter 7: The future of genomics

D. Cyranoski (2015). Ethics of embryo editing divides scientists. *Nature*. 519, 272.

Y. Erlich and D. Zielinski (2017). DNA Fountain enables a robust and efficient storage architecture. *Science*, 355: 950–4.

C.A. Hutchison III et al. (2016). Design and synthesis of a minimal bacterial genome. *Science*, 351: aad6253.

J. Nielsen and J.D. Keasling (2016). Engineering cellular metabolism. *Cell*, 164: 1185–97.

P.E.M. Purnick and R. Weiss (2009). The second wave of synthetic biology: from modules to systems. *Nature Reviews Molecular Cell Biology*, 10: 410–22.

B. Shapiro (2016). Pathways to de-extinction: how close can we get to resurrection of an extinct species? *Functional Ecology*, 31: 996–1002.

L. Tang et al. (2017). CRISPR/Cas9-mediated gene editing in human zygotes using Cas9 protein. *Molecular Genetics and Genomics*, 292: 525–33.

S.A. Zimov (2005). Pleistocene Park: return of the mammoth's ecosystem. *Science*, 308:796–8.

Further reading

The following books (listed in alphabetical order) explore the past, present, and future of genomics. Most are popular science books; many expand on the topics discussed in Chapter 7, including synthetic biology, resurrection biology, genome editing, and personalized medicine.

S. Armstrong (2014). *P53: the gene that cracked the cancer code*. Bloomsbury Sigma.

G. Church and E. Regis (2012). *Regenesis: how synthetic biology will reinvent nature and ourselves*. Basic Books.

J.A. Doudna and S.H. Sternberg (2017). *A crack in creation: gene editing and the unthinkable power to control evolution*. Houghton Mifflin Harcourt.

D. Field and N. Davies (2015). *Biocode: the new age of genomics*. Oxford University Press.

A.M. Lesk (2017). *Introduction to genomics*, third edition. Oxford University Press.

V.K. McElheny (2010). *Drawing the map of life: inside the Human Genome Project*. Basic Books.

S. Pääbo (2015). *Neanderthal man*. Princeton University Press.

J. Parrington (2015). *The deeper genome*. Oxford University Press.

J. Quackenbush (2011). *The human genome: the book of essential knowledge*. Imagine Publishing.

B. Shapiro (2015). *How to clone a mammoth: the science of de-extinction*. Princeton University Press.

J. Shreeve (2004). *The genome war: how Craig Venter tried to capture the code of life and save the world*. Alfred A. Knopf.

M. Snyder (2016). *Genomics and personalized medicine: what everyone needs to know*. Oxford University Press.

Index

cancer (*cont.*)
    passenger mutations and 57–8
    tumour protein 53 (p53) 56
    tumour suppressor genes in
      55–6
Cas9 protein *see* CRISPR
Celera Corporation (Celera
    Genomics) 18, 44
chemical sequencing 14
cloning (molecular) 29, 32–4,
    93–4, 105
cloning (animal) 119
chloroplast 2–3, 82, 85–6
chromosomes 5
Church, George 111–12, 120
codon 6–7
complementary DNA (cDNA)
    34–7
contig 31–2
copy number variation (CNV)
    41–2
Crick, Francis 59–60
CRISPR (clustered regularly
    interspaced short palindromic
    repeats) 114–15, 119–20, 123
cyanobacteria 2, 82, 85
cytochrome c 60–1
cytosine 4–5, 122

D

de-extinction genomics (resurrection
    genomics) 118–21
deoxynucleotides (dNTPs) 115, 117
dideoxynucleotides (ddNTPs)
    115, 117
DNA (deoxyribonucleic acid)
    amplification of 66–7, 88–91,
      93–4
    ancient *see* ancient DNA
      research
    structure of 4–5
DNA Fountain 112
DNA polymerase 15–17
Dolly the sheep 119

E

electrophoresis 14–16
Encyclopedia of DNA Elements
    (ENCODE) project 52–3
endosymbiosis
    mitochondria, chloroplasts
      and 2–3, 82–5
    secondary and tertiary
      endosymbiosis 85–6
environmental genomics
    see metagenomics
environmental PCR 88–91
Escherichia coli 2, 7–8, 86–7
epigenomics 122
Eubacteria *see* Bacteria
eukaryotes (Eukaryota)
    cell structure and diversity
      2–3, 82
    horizontal gene transfer and 86
    origin and evolution of 82–5
eukaryotic signature proteins
    (ESPs) 83–4
exons 8–9, 34, 36, 50
expressed sequence tags (ESTs)
    see transcriptomics
expression vector 105

G

gel electrophoresis 14–16
gene (definition of) 1, 5–7
genetic code (universal genetic
    code) 6–7
genome
    assembly of 30–4
    coverage of 29–31, 41–2
    definition of 1
    draft sequence of 32
    editing of 113–15, 119–20
    hierarchical sequencing of 30–3
    physical map of 33, 44–5
    re-sequencing of 40–3, 54
    shotgun sequencing of 29–33
    whole-exome sequencing of 41

AUTISM
A Very Short Introduction
Uta Frith

This *Very Short Introduction* offers a clear statement on what is
currently known about autism and Asperger syndrome.
Explaining the vast array of different conditions that hide behind
these two labels, and looking at symptoms from the full spectrum
of autistic disorders, it explores the possible causes for the
apparent rise in autism and also evaluates the links with
neuroscience, psychology, brain development, genetics, and
environmental causes including MMR and Thimerosal. This
short, authoritative, and accessible book also explores the
psychology behind social impairment and savantism and sheds
light on what it is like to live inside the mind of the sufferer.

EPIDEMIOLOGY
A Very Short Introduction
Rodolfo Saracci

Epidemiology has had an impact on many areas of medicine;
and lung cancer, to the origin and spread of new epidemics.
and lung cancer, to the origin and spread of new epidemics.
However, it is often poorly understood, largely due to
misrepresentations in the media. In this *Very Short Introduction*
Rodolfo Saracci dispels some of the myths surrounding the
study of epidemiology. He provides a general explanation of
the principles behind clinical trials, and explains the nature of
basic statistics concerning disease. He also looks at the ethical
and political issues related to obtaining and using information
concerning patients, and trials involving placebos.

Science and
Religion
A Very Short Introduction
Thomas Dixon

The debate between science and religion is never out of the news: emotions run high, fuelled by polemical bestsellers and, at the other end of the spectrum, high-profile campaigns to teach 'Intelligent Design' in schools. Yet there is much more to the debate than the clash of these extremes. As Thomas Dixon shows in this balanced and thought-provoking introduction, many have seen harmony rather than conflict between faith and science. He explores not only the key philosophical questions that underlie the debate, but also the social, political, and ethical contexts that have made 'science and religion' such a fraught and interesting topic in the modern world, offering perspectives from non-Christian religions and examples from across the physical, biological, and social sciences.

'A rich introductory text . . . on the study of relations of science and religion.'

R. P. Whaite, Metascience